SpringerWienNewYork

# CISM COURSES AND LECTURES



The series presents lecture notes, monographs, edited works and proceedings in the field of Mechanics, Engineering, Computer Science and Applied Mathematics.
Purpose of the series is to make known in the international scientific and technical community results obtained in some of the activities organized by CISM, the International Centre for Mechanical Sciences.

INTERNATIONAL CENTRE FOR MECHANICAL SCIENCES

COURSES AND LECTURES - No. 478

# MICROSYSTEMS MECHANICAL DESIGN

EDITED BY

FRANCESCO DE BONA
UNIVERSITY OF UDINE, ITALY

ENIKO T. ENIKOV
UNIVERSITY OF ARIZONA, USA

SpringerWienNewYork

The publication of this volume was co-sponsored and co-financed by the UNESCO Venice Office - Regional Bureau for Science in Europe (ROSTE) and its content corresponds to a CISM Advanced Course supported by the same UNESCO Regional Bureau.

This volume contains 172 illustrations


Printed in Italy
SPIN 11817536

In order to make this volume available as economically and as rapidly as possible the authors' typescripts have been reproduced in their original forms. This method unfortunately has its typographical limitations but it is hoped that they in no way distract the reader.

ISBN-10 3-211-37453-1 SpringerWienNewYork
ISBN-13 978-3-211-37453-5 SpringerWienNewYork

# *PREFACE*

*Nowadays, micromechanics (i.e., mechanics of microsystems) is probably one of the most promising and rapidly growing fields among new emerging technologies. In fact, the possibility of reducing the size of mechanical structures to the micro-domain opens a wide variety of possible applications in the biomedical, aeronautical, and automotive fields, in robotics, in molecular engineering, in fiber optics, and in fluidics technology. One of the main aspects that slows down the development of innovative industrial products based on microsystem technology is the existing lack of engineering tools to allow a reliable design of microsystems .*

*The aim of this book is that of collecting the texts o the lectures given at the CISM course on: Microsystems Mechanical Design, hold in July 2004.*

*The purpose of this course was to introduce the basic tools used in the mechanical design of microsystems, the fabrication methods for these systems, and several applications of this technology. The links between micro- and nanotechnologies were also discussed and light was shed on the potential applications of microsystems to nano-scale manipulation of matter, thus introducing the topic of nano-scale engineering mechanics, which will be fully explored in a future course.*

*This book is arranged in 8 sections. In the first section an introduction on microsystems and the techniques for their fabrication will be presented, with a thoroughly description of surface and bulk micromachining techniques and of other microfabrication processes as LIGA and anoding bonding.*

*In the next three sections the case of microsystems loaded electrostatically is considered. In particular scale effects are discussed, the static and the dynamic behaviors of a single degree of freedom electromechanical microsystem are considered; an extension to multi-degree of freedom system is also proposed with the aim analyzing the case of continuum structures discretised with FEM and BEM techniques. The case of a cantilever beam loaded electrostatically was finally considered and an overview of the main analytical and numerical solutions available in literature is proposed.*

*As an example of application, in the fifth section the design case of a microactuator with integrated position sensor was considered, particularly concerning structural and thermal analysis and also the electrical circuit design to achieve the capacitive readout of position.*

*A section is then dedicated to micro and nano assembly; different microassembly technique and devices as mechanical and vacuum grippers, electrostatic and magnetic manipulation approach are discussed. New emerging nanoassembly techniques based on charge writing, and liquid phase and gas phase assembly of nanoparticles are also presented*

*Compliant microstructures are widely used in the design of microsystems. A broad overview of the topics related to the mechanical design of compliant micromechanisms is thus presented in the next section. Design methodologies to be used in the design of devices based on leaf springs, flexural notches and continuum structures with distributed compliance are given, and a critical presentation of the peculiarities of these solutions is provided.*

*In the last section the basic concepts related to the newly emerging field of microfluidics are presented. Following a brief introduction of the general conservation and particular laws, three size effects are introduced. The velocity slip boundary condition for gas flows as well as the electrokinetic and polar-mechanics effects for liquid flow in microdomains are introduced. Pressure-driven gas flows and electrokinetically-driven liquid flows in microchannels are analyzed in details. Finally, several flow diagnostic techniques and fabrication of microfluidic systems are described.*

*In conclusion we thank Professors Velarde and Stepan for their aid and encouragement in the organization of the workshop. We wish also to thank Ms. P. Agnola and all the local organizers in Udine for their hospitality and help in organizing this course.*

*Eniko T. Enikov*
*Francesco De Bona*

## CONTENTS

Preface

Introduction to Micro-Systems and to the Techniques for Their Fabrication
*by E.T. Enikov* ........ 1

Microstructures Under Electrostatic Loads: Discrete System Modelling
*by F. De Bona* ........ 37

Dynamics of Mechatronic Systems at Microscale
*by E. Brusa* ........ 57

Continuum Microstructures Loaded Electrostatically
*by F. De Bona, M. Gh. Munteanu* ........ 81

Design of Electro-Thermal Micro-Positioners: Mechanics and Electronic Position Detection
*by K.V. Lazarov E.T. Enikov* ........ 101

Design of Compliant Micromechanisms
*by F. De Bona, S. Zelenika* ........ 119

Micro and Nano Assemby and Manipulation Techniques for MEMS
*by E.T. Enikov* ........ 135

Microfluidics: Device Science and Technology
*By Lap Man Lee, Luthur Siu Lun Cheung and Yitshak Zohar* ........ 157

# Introduction to Micro-Systems and to the Techniques for Their Fabrication

Eniko T. Enikov

Department of Aerospace and Mechanical Engineering, University of Arizona, Tucson, AZ, USA

**Abstract** This chapter presents an introduction to the main manufacturing processes for micro-systems. After a brief historical overview of the field of micro-systems, their main manufacturing techniques are reviewed. These include substrate preparation, photolithography, oxidation and diffusion, thin-film deposition, and wet and dry etching techniques. MEMS-specific processes including bulk- and surface-micromachining, LIGA, soft-lithography, and anodic bonding are also presented. The material is illustrated with multiple examples of process parameter calculation. Examples of MEMS devices fabricated in the laboratory of the author are also provided.

## 1 Definition of Micro-Electromechanical Systems

The term micro-electromechanical systems (MEMS) was coined to describe a sub-millimeter integrated electro-mechanical system that contains both electrical and mechanical components with sizes in the range of 1 $\mu$m to 1 mm and is fabricated in a massively parallel manner through photolithography. Initially these were simple electrostatically driven cantilevers, fabricated using semiconductor processing techniques (Peterson, 1982). A defining feature of MEMS is their massively parallel manufacture, which results from the use of photolithographic tools for their fabrication. As a result, large quantities of identical planar devices can be made at a very low unit-cost. Another important consequence of the use of photolithography is that the smallest feature producible with this technology has a characteristic dimension roughly equal to the wavelength of light (250-300 nm for UV systems). With the advance of exposure systems that now operate in deep UV, or use even shorter wavelengths such as e-beam lithography and x-ray lithography, the lowest size limit is constantly being broken and currently structures in the 30-100 nm range are possible. Figure 1 illustrates the size definition of MEMS in comparison with other commonly known structures and technologies.

## 2 Brief History of Micro-Electromechanical Systems

The technological origins of MEMS devices can be traced back to 1947, when the first semiconductor transistor was invented in Bell Telephone Laboratories. Approximately

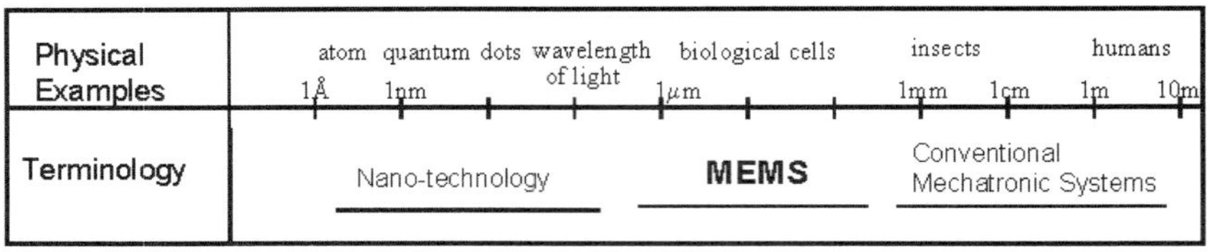


**Figure 1.** The scale of micro devices.

ten years later the first integrated circuit using a Ge sliver attached to a glass slide was demonstrated by Kilby (1964) of Texas Instruments (see Fig. 2).

**Figure 2.** First Integrated Circuit. Courtesy of Texas Instruments

Soon, Robert Noyce of Fairchild Semiconductor announced the development of a planar process capable of integrating multiple transistors and resistors on a single Si substrate, which led to a technological revolution in the electronic industry. Since the 1970s the number of integrated components per unit area has doubled every two to three years, a phenomenon dubbed Moore's law. Elements of micro-electromechanical systems were present even in the early days of the IC industry, though the term MEMS was not coined until the early 1980s. The defining feature of MEMS is the presence of a mechanical component in addition to the electronic elements comprising the system. The first such devices were a resonant gate field effect transistor (RGT) (Nathanson and Wickstrom, 1965) and a pressure sensor based on the piezoresitive effect of Si as demonstrated by Kurtz and Goodman (1974) during the period 1961-70. Soon, new manufacturing techniques such as isotropic and anisotropic etching of Si emerged (Waggener et al., 1967), allowing pressure sensor diaphragms and cavities to be built directly into the Si substrate in a single etching step. The period 1970-80 resulted in the demonstration of the first micro-machined accelerometer, the ink-jet printer nozzle, and various other solid-state sensors. Howe (1980) demonstrated the use polycrystalline silicon layers, forming capacitive structures on the surface of a Si substrate. Soon, the first surface micro-machined resonant cantilever gas sensor was demonstrated by Howe and Muller (1986) at Berkley. The ensuing interest in this technology soon led to an industrial and scientific boom, resulting in polysilicon accelerometers, electrostatic motors, and lateral comb drives. A series of conferences held in 1987-88 led to the widespread acceptance of the terms MEMS, micro-systems, and micro-machines in US, Europe and Japan, respectively. By the early to mid 1990s, polysilicon structures were being used to develop integrated accelerometers (Analog Devices; 1993), light modulators and gratings (Silicon Light Machines; 1992), digital mirror display (Texas Instruments; 1993).

With the maturation of this technology and the emergence of high aspect ratio micromachining methods such as LIGA (Menz, 1992), HEXSIL (Keller and Howe, 1995), soft-LIGA (Sadler et al., 2001), and Deep Reactive Ion Etching (DRIE)(Jansen et al., 2001), MEMS grew closer and closer to their macroscopic counterparts by extending in the third dimension. This development was quite important, since unlike integrated circuits, MEMS are required to perform some sort of mechanical action for which a robust actuator is needed. Interest in applying MEMS to biological and optical applications led to two new terms, "bio-MEMS" and "MOEMS" (Micro-Optical Electro-Mechanical Systems). Numerous foundries were opened in the US (e.g., MCNC and Sandia National Labs) and across the world to provide services to the MEMS community. In 1998-2002, coinciding with the "dot-com" (.com) era of economic boom and bust, many companies invested in optical MEMS for telecommunications and optical switching. Unfortunately, this enthusiasm did not pay out and many companies were forced to close their doors. Since 2000, the research focus has shifted to radio-frequency (RF) MEMS for steerable antennas, switches, and variable RF attenuators and resonators. The interest in bio-MEMS and lab-on-a-chip applications continues to be strong, as is the area of nano-electromechanical systems (NEMS), a term reserved for systems with characteristic lengths below 100 nm.

## 3 Basic MEMS Fabrication Process and Terminology

### 3.1 Substrates

Since MEMS emerged as a spin-off of well-developed semiconductor processes, the most commonly used substrates for MEMS are semiconductors such as Si, Ge, and GaAs. Due to the need for chemically inert materials, glass, ceramics, and lately even printed circuit boards (Enikov and Lazarov, 2003) have been utilized in the fabrication of MEMS devices. Silicon substrates are the most popular, since a wide range of processes and equipment is available for this material. Other commonly used substrates include boro-silicate glasses (Pyrex), alumina, and a range of polymeric materials including polymethyl-methacrylate (PMMA) and poly(dimethylsiloxane) (PDMS) rubber, mostly used in micro-fluidic and biomedical micro-devices.

Silicon, which is a Group IV element, is the most widely used substrate material. Its unit cell can be constructed by starting with a face-centered cubic (FCC) cell and adding four more atoms at locations $(a/4, a/4, a/4)$, $(3a/4, 3a/4, a/4)$, $(3a/4, a/4, 3a/4)$, and $(a/4, 3a/4, 3a/4)$, where $a$ is the lattice constant of the original FCC cell. These four atoms form a diamond lattice within the FCC lattice (see Fig. 3(a)). Large single crystals of Si are grown from a melt in a process called Czochralski growth. The process starts with a seed crystal approximately 0.5 cm in diameter, which is brought in contact with a Si melt. Through a slow withdrawal and rotation, a cylindrical boule with diameter up to 300 mm and length reaching 1-2 m is formed. The boule is then sliced and the slices polished to form very smooth substrates (wafers). The crystal orientation and doping type of each wafer are denoted by standard cuts ("flats") on its periphery, as indicated in Figure 4. The following are typical specifications for a starting Si wafer:

- Orientation: (100) or (110)
- Size: 2", 3", 4", 6", 8"
- Thickness: 11-13 mils, 1 mil=1/1000 of an inch=25.4 $\mu$m
- Dopant type: B, P, As
- Dopant level (resistivity): 3-10 $\Omega$-cm
- Defects: dislocation density less thab 100/cm$^2$
- Bow: typically less than 20 $\mu$m

A thorough substrate cleaning is also required at the beginning of the process to remove any organic films, particulates, native oxides, and metal ion contaminants, and to reform a contamination-free native oxide. Contamination usually occurs during the grinding and polishing process (heavy metals), from human handling, and from other chemicals coming in contact with the wafer or processing equipment. Therefore MEMS, similar to IC chips, are manufactured in clean-room environments, where the number of airborne particles is controlled.

Similar to integrated circuit processing technology, MEMS are built through multiple cycles of material deposition (additive process), patterning (image formation), and pattern transfer (subtractive process). This cycle is illustrated in Figure 5. Electronic circuits typically use 10-18 photolithographic (patterning) steps, while MEMS structures can be usually constructed with fewer steps. These three fundamental processes are described briefly in the following sections.

### 3.2 Photolithography

Photolithography is the core of almost all fabrication process, since this step defines the geometry of the structures used in a micro-device. There are two types of photosensitive materials (photoresists or simply resists): positive and negative (tone) resists. With positive resists, the areas exposed to ultraviolet (UV) light become soluble through a photolysis process occurring in the polymer network. Conversely, negative resists become insoluble (cross-linked) in the areas exposed to UV light. Positive resists

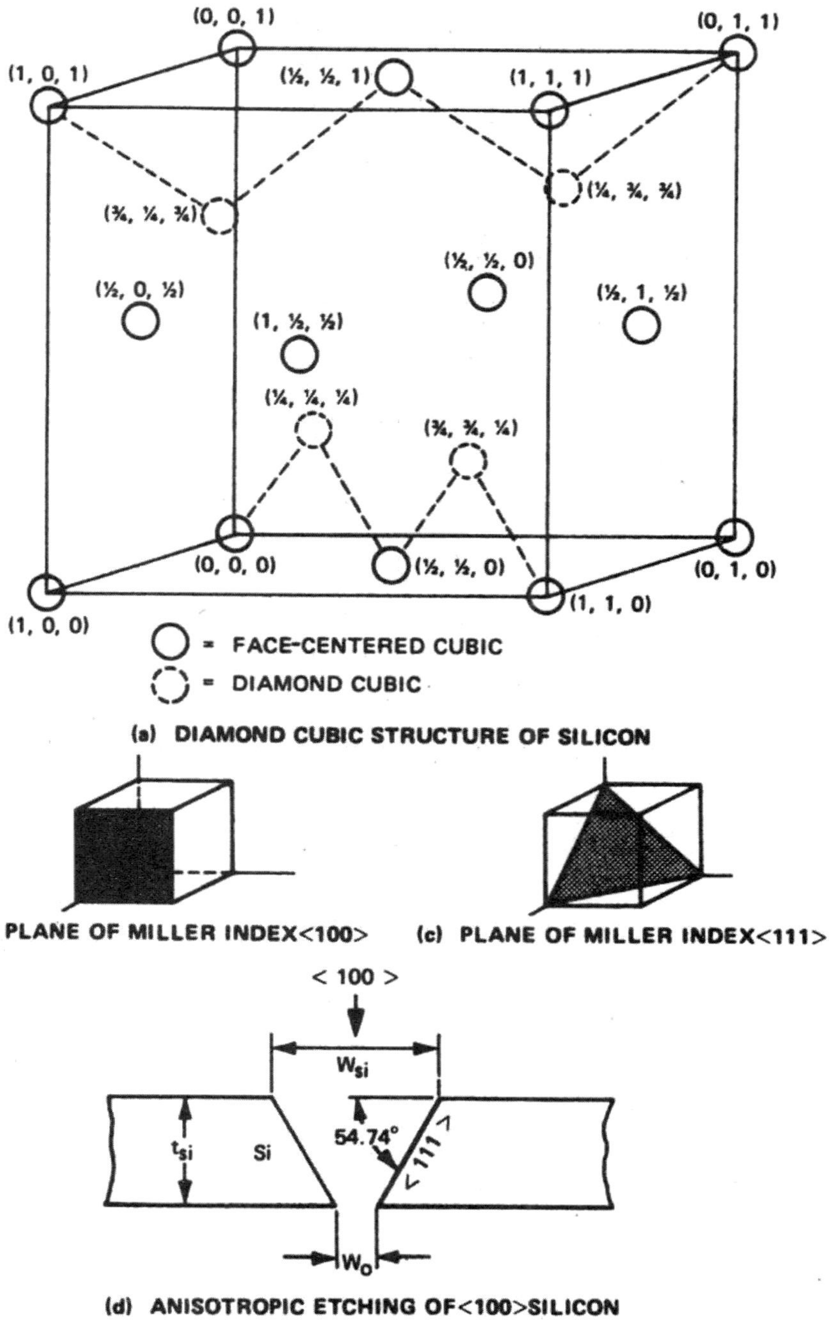


**Figure 3.** (a)Silicon lattice; (b-c) key crystallographic planes; (d) anisotropic etch delineating (111)planes. Reproduced with permission from Brodie and Muray (1992).

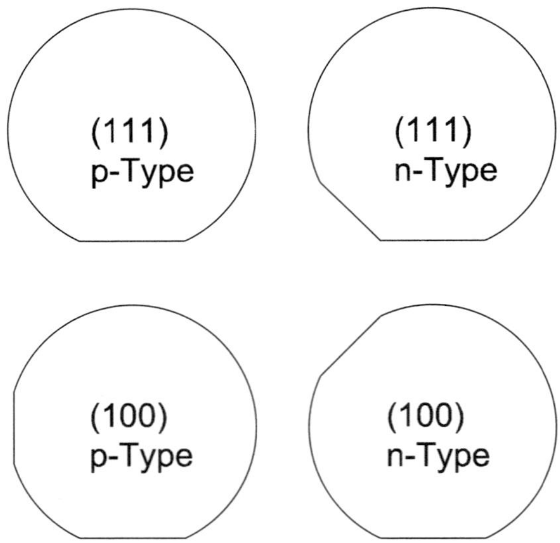


**Figure 4.** Wafer types and flat orientation

are sensitized with diazoquinone (DQ), which is base-insoluble but, when exposed to UV radiation, photolyses and produces carbene, which is then transformed to ketene (using the residual water in the film), and thus becomes soluble in a caustic solution.

Negative resists undergo cross-linking through a photon-initiated radical formation of the azid group, which releases nitrogen.

Differences between positive resist and negative resist pattern formations is shown in Figure 6. As illustrated in the figure, light diffraction and scattering produces wider lines in the negative resist and narrower lines in the positive resist. The following terms are commonly used to characterize the quality of the photolithographic process (ome of these are defined procedurally):

- *Resolution* is the smallest line-width that can be printed reliably under typical variations in the manufacturing conditions.
- *Line-width* is the horizontal distance between the resist-air interface at a given height above the substrate. The defined line-width depends on the method used to measure it. For example, different line-widths will be established though optical, mechanical scanning.
- *Contrast* is the rate of formation or scission of a cross-linked network for negative resists and positive resists, at a constant exposure dose.

The contrast $\gamma$ can be determined experimentally via measurement of the developed

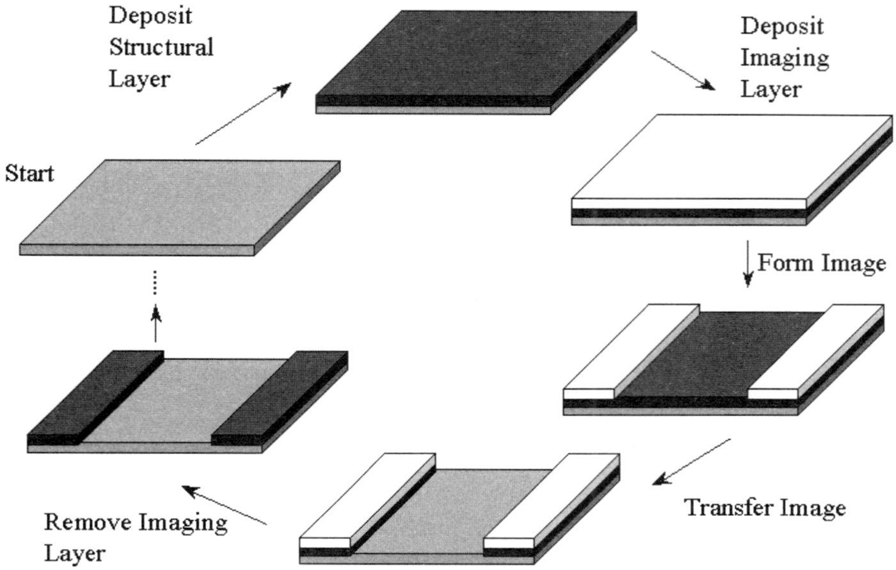


**Figure 5.** MEMS and IC processing cycle.

resist thickness vs. exposure dose (Wolf and Taubert, 2000). For example, for a negative resist the contract $\gamma_n$ is determined from

$$\gamma_n = \log_{10}^{-1} \frac{D_0}{D_i}, \tag{3.1}$$

where $D_i$ is the critical exposure dose under which no cross-linking occurs and $D_0$ is the extrapolated dose for which complete cross-linking occurs (see Fig. 7). For positive photoresists, the contrast value is determined analogously

$$\gamma_p = \log_{10}^{-1} \frac{D_c}{D_0}, \tag{3.2}$$

where $D_c$ is the critical exposure dose under which there is always some undeveloped (cross-linked) photoresist and $D_0$ is the extrapolated dose under which all of the photoresist remains cross-linked. Commonly used positive photoresists have higher contrast values in comparison with negative resists and can therefore produce images with higher resolution. The adhesion to silicon and the chemical resistance of positive resists, however, are somewhat inferior. The main drawback of negative photoresists is their significant swelling during development, which limits the minimum feature size to about 2 $\mu$m.

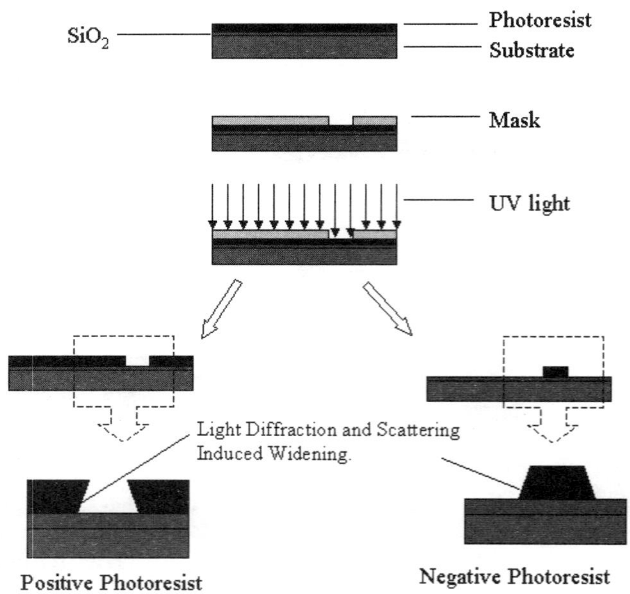


**Figure 6.** Positive and negative resist patterns

Due to light diffraction and scattering, it is clear that the resist thickness is very important in achieving a desired resolution. As a rule of thumb, the minimum feature size cannot be significantly smaller than the thickness of the photoresist. Therefore, methods for producing a uniform film over the entire substrate are needed. Several such techniques are available:

- Spray
- Electrophoretic deposition
- Dip-coating
- Lamination ("dry resists")
- Spin coating

Among these, the most common is spin-coating, which results in very uniform films. In spin-coating, the substrate is spun at a given rate, allowing centrifugal forces to thin the dispensed solution of photoresist to a film with the desired thickness. This relatively simple process results in a remarkably uniform layer, with sub-micron thickness variations over the surface of a 100-200-cm-diameter wafer. A simple insight into the reasons behind this can be gained from the solution of an axisymmetric flow problem of a Newtonian fluid driven by centrifugal forces. Equating the viscous drag in the radial direction with the centrifugal forces one has

$$\eta \frac{\partial v}{\partial z^2} = -\rho \omega^2 r, \tag{3.3}$$

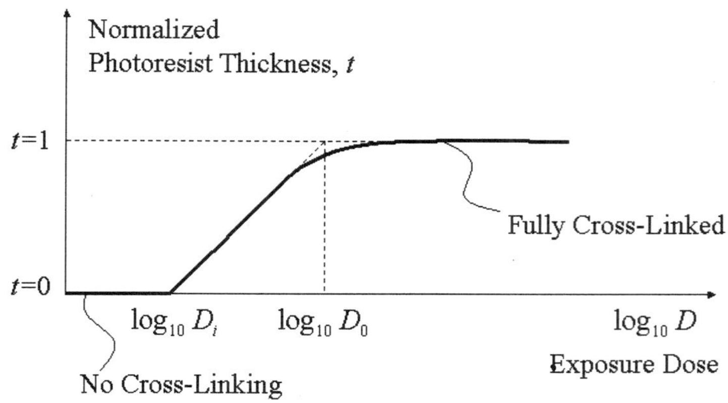


**Figure 7.** Contrast plot of negative photoresist.

where $\eta$ is the resist viscosity, $\rho$ it its density, $v$ is the fluid velocity, and $z$ and $r$ are the axial and radial coordinates. Simple integration with $v(z=0)=0$ and $\partial v/\partial z(z=h)=0$ results in

$$v = \frac{\rho\omega^2 r}{\eta}\left[-\frac{z^2}{2}+hz\right], \tag{3.4}$$

where $h$ is the thickness of the fluid (photoresist). The total mass of fluid flowing per unit length of a circle with radius $r$ is

$$q = \int_0^h v(z,r)dz = \frac{\rho\omega^2 r}{\eta}\frac{h^3}{3}. \tag{3.5}$$

Using mass conservation, one can find that the fluid height at a distance $r$ from the axis of rotation is governed by a simple first-order differential equation

$$\dot{h} = -\frac{1}{r}\frac{\partial}{\partial r}(rq) = -\frac{1}{r}\frac{\partial}{\partial r}\left(\frac{\rho\omega^2 r^2}{\eta}\frac{h^3}{3}\right) \approx -\frac{2}{3}\frac{\rho\omega^2}{\eta}h^3, \tag{3.6}$$

where the height $h$ has been assumed independent from the radius $r$, which is experimentally observed. Solving Eq. (3.6), one finds that the film thickness is inversely proportional to the square of the angular velocity

$$h(t) = \frac{h_0}{\sqrt{1+\frac{4}{3}h_0^2\frac{\rho\omega^2}{\eta}t}}, \tag{3.7}$$

where $h_0$ is the initial film height. For long times, the height is independent of the initial amount of dispensed liquid $h_0$ and is given by

$$h^{\text{inf}} = \left(\frac{4}{3}h_0^2\frac{\rho\omega^2}{\eta}t\right)^{-\frac{1}{2}}, \tag{3.8}$$

which predicts a zero height at infinitely long times. In reality, a finite height is established, which is not equal to the predictions of Eq. (3.8). This is due to solvent evaporation during spinning, which changes the viscosity of the film and arrests the flow after some time. An empirical model describing the residual thickness is given by

$$h^{\text{inf}} = \frac{KC^{\beta}\eta^{\gamma}}{\omega^{\alpha}}, \tag{3.9}$$

where $K$ is a calibration constant, $C$ is the polymer concentration in grams/100 ml solution, $\eta$ is the intrinsic (kinematic) viscosity, and $\omega$ is the "spin speed" in revolutions per minute (rpm). Meyerhofer (1978) has shown that when the evaporation is accounted for, the value of the exponent $\alpha$ is approximately 2/3. Typical spin speeds range from 500 rpm up to 7000 rpm, producing films with a thickness of tens of microns down to a fraction of a micron.

A typical photolithography step includes:

- Spin-coating of photoresist at 1000-4000 rmp.
- Pre-exposure bake (soft bake) at 90-100 °C to remove residual solvent in the film;
- UV exposure at 130-170 mJ/cm$^2$
- Development 1-5 min
- Post-exposure bake at 110-125°C to remove leftover solvents and make the film more resistive to chemical attacks.

The photolithographic step is almost always preceded by a deposition step, which lays down the layer to be patterned. There are many deposition techniques currently in use in the fabrication of MEMS devices. The main ones used for the deposition of films less than 2-3 $\mu$m thick are physical vapor deposition (PVD) and chemical vapor deposition (CVD). Techniques producing thicker layers (above 4-5 $\mu$m) are electrodeposition (electroplating), polymer casting, spray coating, or high-pressure oxidation (HIPOX) of silicon.

### 3.3 Thermal Oxidation of Silicon

Thermally grown $SiO_2$ layers are the most important insulator in semiconductor devices. They serve as the gate insulator in CMOS transistors, dielectric in capacitive elements, as well as a masking material for a variety of process steps such as doping and etching. In micro electro-mechanical systems, the silicon dioxide (oxide) serves similar roles. There are many reasons behind the popularity of this dielectric in semiconductor device fabrication, the most important of which is its good electrical properties, ease of film growth, and its superior masking properties. The interested reader is referred to a monograph by Nicollian and Brews (1982) for an extensive discussion on the electrical properties of silicon dioxide and a variety of methods for its characterization.

The silicon dioxide is most commonly deposited via thermal oxidation in high-temperature quartz tubes under the supply of oxygen, steam, and possibly an inert carrier gas such as nitrogen or argon. Typical oxidation temperatures range from 850 °C to about 1272 °C. Figure 8 shows a cross section of the silicon wafer with a thin silicon dioxide film growing on its surface. As illustrated in the figure, the oxidation takes place at the interface between the single-crystal silicon surface and the already grown film, which requires oxygen to diffuse through the already grown oxide in order to continue the film growth. From a technological point of view, it is important to be able to predict the required time for growing a film with a given thickness. In order to do this, a model based on the diffusion of oxidants (oxygen or water) was developed by Deal and Grove (1965) that allows accurate prediction of the final thickness of the grown film. With the notation from Figure 8, one can set up equilibrium flux conditions for oxygen arriving from the gas phase to the surface, then diffusing through the oxide, and finally reacting with the silicon surface. In the Deal-Grove model, each of these fluxes is related to

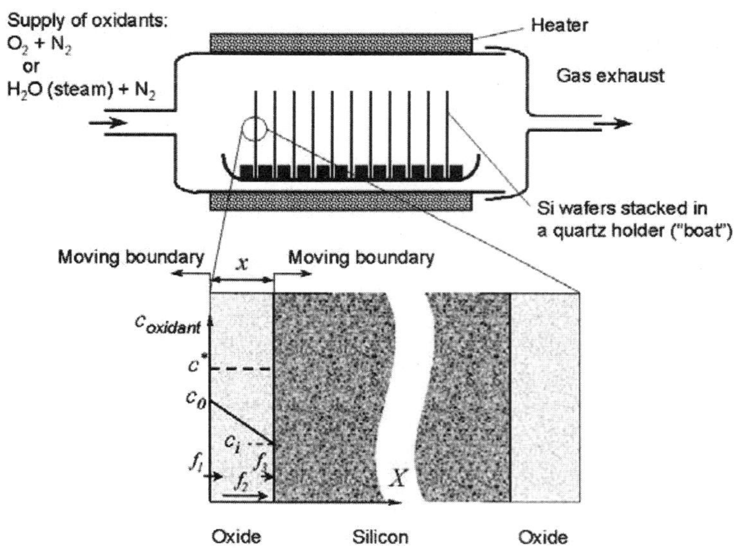


**Figure 8.** Wafer oxidation equipment (upper half); and parameters used in the Deal-Grove's oxidation kinetics model (bottom half).

the concentration of the oxidant diffusing through the silicon oxide. The flux entering the oxide from the gas phase is proportional to the difference between the equilibrium concentration $c^*$ and the actual surface concentration $c_o$ at the oxide-air interface

$$f_1 = h(c^* - c_o), \tag{3.10}$$

where $h$ is the mass transfer coefficient for the air-oxide interface. The equilibrium concentration $c^*$ is proportional to the external gas pressure through Henry's law

$$c^* = HP, \tag{3.11}$$

where $P$ is the partial pressure of the oxidant in the chamber and $h$ is the Henry's law coefficient. Assuming a linear concentration profile, this flux is given by

$$f_2 = \frac{D(c_o - c_i)}{x}. \tag{3.12}$$

Finally, the oxidant is consumed by the silicon oxidation reaction

$$f_3 = Kc_i, \tag{3.13}$$

where $K$ is the reaction rate constant. Combining Eqs. (3.10), (3.12), and (3.13) and eliminating the variable $c_o$ results in

$$c_i = \frac{c^*}{1 + K/h + Kx/D}. \tag{3.14}$$

A differential equation describing the growth of the oxide thickness, $x(t)$ is derived by using Eq. (3.13), and realizing that the rate of change of the consumption of the oxidant is proportional to the growth rate

$$N_i \frac{dx}{dt} = f_3 = Kc_i = \frac{Kc^*}{1 + K/h + Kx/D}, \tag{3.15}$$

where $N_i$ is the number of oxidant molecules needed to grow a unit volume of oxide. For $SiO_2$, $N_i = 2.2 \times 10^{22}$ molecules/cm$^3$ (Brodie and Muray, 1992). Integrating Eq. (3.15) between the initial and final oxide thickness $x_i$ and $x_o$, respectively, for the time interval $[0, t]$ gives

$$x_o^2 + Ax_o = B(t + \tau), \tag{3.16}$$

where

$$\begin{aligned} A &= 2D\left(\frac{1}{h} + \frac{1}{K}\right) \\ B &= \frac{2Dc^*}{N_i} = \frac{2DHP}{N_i} \\ \tau &= \frac{x_i^2 + Ax_i}{B}. \end{aligned} \tag{3.17}$$

The parameter $\tau$ measures the effect of the initial oxide thickness, which affects the rate of oxidation. Equation (3.16) has one positive root

$$x_o = \frac{A}{2}\left(\sqrt{1 + \frac{t + \tau}{A^2/4B}} - 1\right). \tag{3.18}$$

There are two growth regimes. For short oxidation times and thin initial oxides ($t + \tau \ll A^2/4B$), the growth is approximately linear

$$x_o \approx \frac{B}{A}(t + \tau). \tag{3.19}$$

It can be noted that $B/A$ is dependent on the reaction rate $K$, but is independent from the diffusion constant. This regime is therefore known as reaction-limited growth. If a longer oxidation is performed and/or there is a significant initial oxide ($t + \tau \gg A^2/4B$), the growth is parabolic

$$x_o \approx \sqrt{B(t + \tau)}. \tag{3.20}$$

In this regime, the growth rate is parabolic and is dependent on $B$, which is a function of the diffusion constant, but is independent from the reaction rate. In both cases, the growth coefficients are strongly dependent on the equilibrium concentration $c^*$ and, through (3.11), on the external pressure. The growth rate can be increased dramatically by increasing the external pressure or the oxidation temperature. Since practical temperature limits are around 1200 °C, oxides beyond 2 $\mu$m are grown using high-pressure oxidation (HIPOX) at 10-20 atm. When water vapor (steam) is used instead of oxygen, the growth rate is usually higher due to the higher equilibrium concentration of water, $c^*$, in the oxide (higher solubility). This is commonly used in MEMS fabrication, where thick oxides are grown in steam ("wet" oxide) to be used as etch or diffusion masks. When a high-quality oxide is needed, as in the gates of field effect transistors, steam use is avoided to ensure a denser oxide with fewer electrical traps (defects). The parameters $A$ and $B$ have an Arrhenius-type dependence on temperature, as well as the nature of the oxidant ($O_2$ vs. $H_2O$). The Deal-Grove model (Deal and Grove, 1965) does not accurately model the oxide growth for very thin oxides (below 30 nm), where some other oxidation mechanisms appear to be at play. To correct for these effects, a time offset parameter, $\tau$, is used to correct for the apparently thin initial oxide when using Eq. (3.19) for thin, dry oxides. Table 1 lists the experimentally determined values for several temperatures and oxidation conditions, along with the values of this empirically established offset parameter.

**Table 1.** Oxidation rate constants, after Deal and Grove (1965).

| | Wet Oxide | | Dry Oxide | | |
|---|---|---|---|---|---|
| Temperature [°C] | A [$\mu$m] | B [$\mu m^2$/hr] | A [$\mu$m] | B [$\mu m^2$/hr] | $\tau$ [hr] |
| 920 | 0.50 | 0.203 | 0.235 | 0.0049 | 1.4 |
| 1000 | 0.226 | 0.287 | 0.165 | 0.0117 | 0.37 |
| 1100 | 0.11 | 0.510 | 0.090 | 0.027 | 0.076 |

### 3.4 Doping

Doping, one of the most essential processes used in creating integrated circuits, introduces electrically active impurities, resulting in local changes of conductivity and the formation of passive and active devices. In micro-system fabrication, doping is used for similar purposes, as well as to modify the electro-chemical properties of Si. For example, a heavily doped p+ silicon is resistant to alkaline etching and can be used as an electro-chemical etch stop. This process can be utilized to produce Si diaphragms in pressure sensors or the nozzles of inkjet printer-heads (Brodie and Muray, 1992). The process is illustrated in Figure 9.

Pressure-sensitive resistors (piezo-resistors) integrated into the structure can also be fabricated through doping. These resistors are typically used to measure the stress in a diaphragm (pressure sensor) or micro-cantilver force sensors. There are two main methods for introducing impurities into semiconductors: diffusion and ion-implantation.

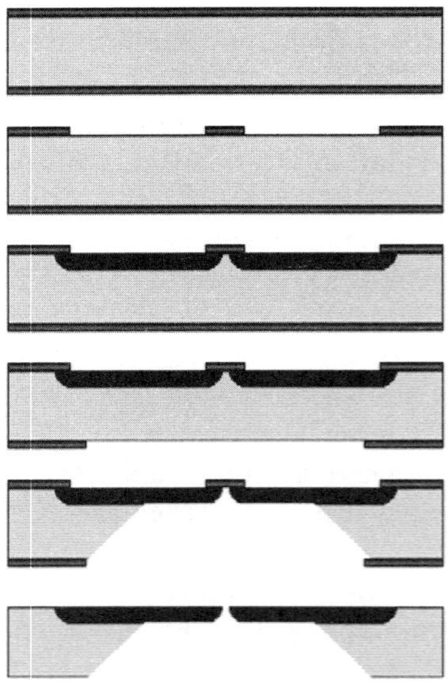


**Figure 9.** Fabrication sequence of a micro-nozzle for a printer-head via heavy (p+) doping, followed by anisotropic etching.

In the case of diffusion, impurities (dopants) are introduced from a solid, gas, or liquid source and then diffused into the substrate in what is known as a classical two-step thermal diffusion process. During the first step (predeposition), the total amount of impurities (dose) is established by maintaining the concentration at the surface at the solid solubility limit. During the second step (drive-in), thermal diffusion under no additional supply of surface impurities re-distributes them into the substrate to establish the desired metallurgical junction depth. During each step, the dopant distribution, the junction depth, and, subsequently, the resistance of the diffused resistor can be determined from the (1D) diffusion equation

$$\frac{\partial c}{\partial t} = D\frac{\partial^2 c}{\partial x^2}, \tag{3.21}$$

where

$$D = D_0 \exp(-E_a/kT) \tag{3.22}$$

is a temperature-dependent diffusion coefficient with activation energy $E_a$. Strictly speaking, the parameter $D_0$ is, in fact, dependent on the impurity type and its concentration. For example, the diffusion coefficient of boron (a p-type dopant) has the form (Runyan and Bean, 1990)

$$D = \left(D^* + D^+\frac{p}{n_i}\right)\exp(-E_a/kT) = \left(0.037 + 0.41\frac{p}{n_i}\right)\exp(-3.46/kT); \quad (\mathrm{cm}^2/s), \tag{3.23}$$

where $p$ is the majority carrier concentration, $n_i$ is the intrinsic carrier concentration at the process temperature, and $E_a$ is the activation energy in eV. Due to the high process temperature, the intrinsic carrier concentration is quite high. In this case, the carrier concentration for the p-type semiconductor is given by (Sze, 1981)

$$p = \frac{c_A - c_D}{2} + \left(\frac{(c_A - c_D)^2}{4} + n_i^2\right)^{1/2}, \tag{3.24}$$

where $c_A$ and $c_D$ are the acceptor and donor concentrations, respectively. A similar expression is valid for n-type materials. An empirical expression for the intrinsic carrier concentration of Si was developed by Morin and Maita (1954)

$$n_i^2 = 1.5 \times 10^{33} \exp(-1.21/kT), \tag{3.25}$$

which allows determination of the carrier concentration at the process temperature. Using Eqs. (3.23)-(3.25), the diffusion coefficient for the dopand of interest (boron) can be estimated. For example at 900 °C, assuming that the boron concentration is at its solid solubility limit, $c_A = c_s = 1.1 \times 10^{20}$ cm$^{-3}$, and the substrate is an n-type Si with background concentration of $c_D = 4 \times 10^{15}$ cm$^{-3}$, the boron diffusion coefficient is estimated to be $D = 1.56 \times 10^{-14}$ cm$^2$/s.

With a sufficiently accurate estimate of the diffusion coefficient, Eq. (3.21) can be solved subject to different boundary conditions during each of the two doping steps. The predeposition is performed with excess dopants, which maintains the surface concentration equal to the solid solubility limit $c_s$ at the process temperature

$$\begin{aligned} c(t,0) &= c_s \\ c(t,\infty) &= 0. \end{aligned} \tag{3.26}$$

The corresponding solution is described by the complementary error function (Ghez, 2001)

$$c(t,x) = c_s \mathrm{erfc}(\frac{x}{2\sqrt{Dt}}), \tag{3.27}$$

where

$$\mathrm{erfc}(x) = \frac{2}{\sqrt{\pi}} \int_x^\infty \exp(-s^2) ds.$$

The total dose introduced into the substrate during predeposition is then

$$Q = \int_0^\infty c(t,x) = 2c_s\sqrt{Dt/\pi}. \tag{3.28}$$

For example, using the boron diffusion data at 900 °C, the total dose of dopant introduced over a diffusion time of 17 minutes ($t = 1020$ s) is $Q = 5 \times 10^{14}$ cm$^{-2}$. During the drive-in step, no additional impurities are introduced. Instead, the dopants are gradually driven into the substrate, thus increasing the junction depth and reducing the peak surface concentration. The corresponding boundary conditions in this case are

$$\frac{\partial c(t,0)}{\partial x} = 0$$

$$c(t, \infty) \quad = \quad 0. \tag{3.29}$$

The concentration profile in this case is approximated by a Gaussian distribution

$$c(t, x) = \frac{Q}{\sqrt{\pi D t}} \exp\left(\frac{-x^2}{4Dt}\right), \tag{3.30}$$

which is a good approximation as long as the diffusion length of the second step is significantly larger than that of the first step, i.e., $\sqrt{D_2 t_2} \gg \sqrt{D_1 t_1}$. The location where the dopant concentration is equal to the background concentration, $c_b$, is known as the metallurgical junction depth. It can be determined from Eqs. (3.27) and (3.30) for the two types of solutions, respectively

$$c(t, x_j) = c_b \Rightarrow x_j = \begin{cases} 2\sqrt{Dt}\ \mathrm{erfc}^{-1}(c_b/c_s) & \text{for erfc() distribution,} \\ 2\sqrt{Dt \ln\left(\frac{Q}{\sqrt{\pi D t}}/c_b\right)} & \text{for Gaussian distribution.} \end{cases} \tag{3.31}$$

Using the junction depth, the dopant distribution, and the mobility of the majority carrier, one can determine the so-called sheet-resistance, $R_{\Box}$, which is a measure of the resistivity and allows a convenient calculation of the total resistance of the diffused resistor. The sheet-resistance is defined as the average resistivity of the layer divided by the junction depth

$$R_{\Box} = \frac{\bar{\rho}}{x_j} = \frac{1}{\int_0^{x_j} \sigma(x) dx}, \tag{3.32}$$

where $\sigma(s)$ is the total conductivity. Using the electron and hole concentrations, $n$ and $p$, and mobility, $\mu_n$ and $\mu_p$, respectively, the total conductivity can be determined through

$$\sigma = q(\mu_n n + \mu_p p), \tag{3.33}$$

where $q = 1.6 \times 10^{-19}$ C is the electron's charge. An empirical relation between the electron and hole mobilities and their concentrations was developed by Caughey and Thomas (1967),

$$\mu = \mu_{\min} + \frac{\mu_{\max}}{1 + \left(\frac{c}{c_{\text{ref}}}\right)^{\alpha}}, \tag{3.34}$$

where the values of the parameters are listed in Table 2.

Using Eq. (3.34) in Eq. (3.32), one can determine the average conductivity and sheet resistance of the diffused resistor. For the n-type Si substrate with background concentration of $c_D = 4 \times 10^{15}$ cm$^{-3}$, assuming that only pre-deposition of boron is performed for 17 minutes at 900 °C, the junction depth is $x_j = 0.24$ $\mu$m and the resulting sheet resistance is $R_{\Box} = 222$ $\Omega/\Box$. The total resistance of a $W = 20$-$\mu$m-wide and $L = 200$-$\mu$m-long resistor made with this process is

$$R = R_{\Box} \frac{W}{L} = 2220\ \Omega. \tag{3.35}$$

The analysis above is used in the design of strain-sensitive resistors used in pressure- or micro-cantilever sensors. A typical structure of a micro-mechanical cantilever sensor

**Table 2.** Mobility parameters for electrons and holes from (Caughey and Thomas, 1967).

| Parameter | Electrons (As, Sb, P) | Holes (B,Ga,Al) |
|---|---|---|
| $\mu_{\min}$ [cm$^2$/Vs] | 65 | 47.7 |
| $\mu_{\max}$ [cm$^2$/Vs] | 1330 | 495 |
| $c_{\text{ref}}$ [atoms/cm$^3$] | $8.5 \times 10^{16}$ | $6.3 \times 10^{16}$ |
| $\alpha$ | 0.72 | 0.76 |

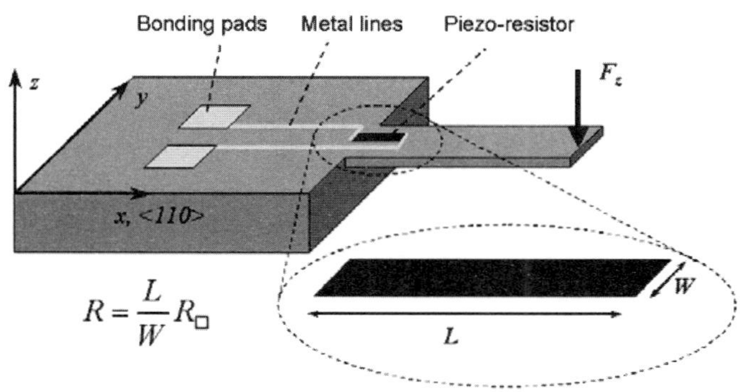


**Figure 10.** Micro-cantilever force sensor with diffused or implanted piezo-resistor.

consisting of a resistor at the base of an n-type Si cantilever is shown in Figure 10. The cantilever and the piezo-resistor are oriented along the $< 110 >$ direction. The resistance change as a function of the two principal stresses, $\sigma_x$ and $\sigma_y$, is given by the piezo-resistive coefficient $\Pi_{44}$

$$\frac{\Delta R}{R} = \frac{\Pi_{44}}{2}(\sigma_x - \sigma_y), \tag{3.36}$$

where $\Pi_{44} \approx 80 \times 10^{-11}$ Pa$^{-1}$ is the piezo-resistive coefficient of Si. Very often, in order to reduce the temperature sensitivity of the fabricated sensor, multiple resistors are used in a Wheatstone bridge configuration. Such a pressure sensor, developed in our laboratory, is shown in Figure 11, where two of the resistors are placed close to the outer edge of the bulging diaphragm while another two are located on the un-deformed portion of the substrate.

### 3.5 Vacuum Deposition Processes

Physical vapor deposition techniques rely on thermal evaporation or sputtering of the desired material and subsequent deposition onto the substrate. Metals are usually

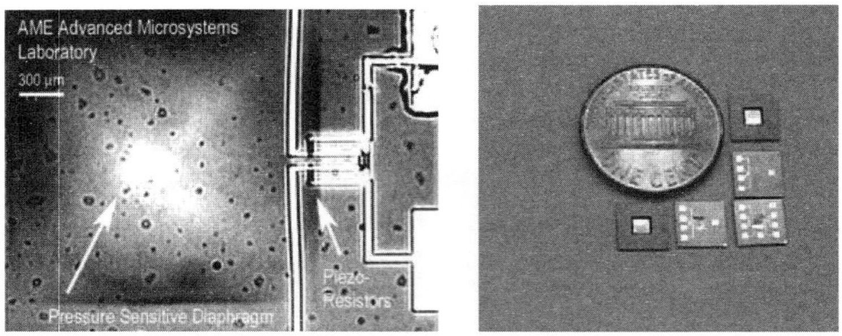


**Figure 11.** Piezo-resistive pressure sensor: diaphragm with Wheatstone bridge (left); unpackaged sensors with Al bonding pads (right).

deposited via evaporation or sputtering while dielectrics and piezoelectric materials are sputtered. The PVD is carried out under a vacuum to avoid oxidation of the film or incorporation of unwanted species. This also ensures intimate contact of the deposited atoms with the semiconductor substrate. An insight into the deposition process can be gained from the kinetic theory of monoatomic gases. The velocities of the molecules of such a gas, $v$, obey the Maxwell distribution

$$P(v) = 4\pi \left[\frac{m}{2\pi kT}\right]^{3/2} v^2 e^{-mv^2/2kT}, \tag{3.37}$$

where $m$ is the mass of the gas molecule and $k$ is the Boltzman factor. First and second moments of this distribution provide the average linear and quadratic velocities

$$\begin{aligned} \bar{v} &= \int_0^{\inf} P(v) v dv = \sqrt{\frac{8kT}{\pi m}}, \\ \bar{v^2} &= \int_0^{\inf} P(v) v^2 dv = \frac{3kT}{m}. \end{aligned} \tag{3.38}$$

From momentum balance, one can show that the pressure of the gas is proportional to the mean square of the velocity, resulting in the well-know ideal gas law

$$p = mn\bar{v_x^2} = \frac{mn\bar{v_2}}{3} = nkT, \tag{3.39}$$

where $n$ is the number of molecules per unit volume. The free mean path of the molecules, $\lambda$, is given by

$$\lambda = \frac{1}{\sqrt(2)\pi d^2 n} = \frac{kT}{\sqrt{2}\pi d^2 p}, \tag{3.40}$$

where $d$ is the atomic (molecular) collision diameter. Finally, the average number of molecules striking a unit surface per unit time is

$$J_x = \frac{n}{2}\bar{v_x} = \frac{n}{2}\frac{\bar{v}}{2} = \frac{p}{\sqrt{2kT\pi m}}. \tag{3.41}$$

When used with a source of area $A$ and vapor pressure $p_e$, after mulitiplication with $Am$, Eq. (3.41) is known as the Langmuir mass evaporation rate

$$R_m = mJ_x = \sqrt{\frac{m}{2\pi kT}} p_e A. \tag{3.42}$$

Using Eq. (3.42), one can verify that reasonable deposition rates are achieved at vapor pressures of 10 mTorr or greater. According to the vapor pressure data for several common metals shown in Figure 12, the temperature needed is around the melting point of most metals. The high reactivity of these metals at these temperatures requires the use of special inert crucibles. If a source of contaminant (water vapor for example) is

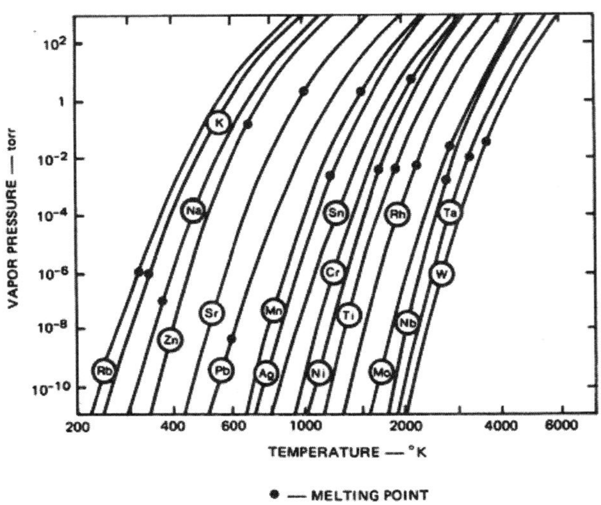


**Figure 12.** Vapor pressures of common metals (reproduced with permission from Brodie and Muray (1992).)

present in the chamber, resulting in a constant base pressure of 1 $\mu$Torr, analysis of Eq. (3.41) results in $4.8 \times 10^{14}$ molecules/cm$^2$-sec. Therefore the commonly used chamber base pressures (prior to heating the source) are below 1$\mu$Torr to prevent incorporation of contaminants into the film. At these pressures, the mean free path (MFP), $\lambda$, of the molecules is 50 m, a distance much larger than the size of the deposition chamber. Thus the molecules travel in a molecular regime, following the line of sight from the source to the target substrate. In other deposition techniques, such as ion (plasma) sputtering, the chamber pressure is approximately 1 mTorr, resulting in a somewhat more randomized (conformal) deposition. Typical values of the MFP at three common pressures are

$$\begin{aligned} \lambda &= 50\ \mu\text{m}, \quad p = 1\ \text{Torr}; \\ \lambda &= 5\ \text{cm}, \quad p = 1\ \text{mTorr}; \\ \lambda &= 50\ \text{m}, \quad p = 1\ \mu\text{Torr}. \end{aligned} \tag{3.43}$$

In the molecular flow regime, the deposition rate $D$ (film thickness growth per unit time) is given by the so-called "cosine law" of vacuum deposition

$$D = \frac{R_m}{\pi \rho R_1^2} \cos\phi_1 \cos\theta_1, \tag{3.44}$$

where $R_1$ is the distance between the source and the target (see Fig. 13). From Eq. (3.44), one can observe that the largest deposition rate is achieved when the substrate is directly above the source. To avoid such non-uniformity, the evaporator chambers contain planetary assemblies, which rotate the wafers on a sphere having the evaporation source on its south pole. Geometrically, this means that

$$\cos\phi_2 = \cos\theta_2 = \frac{r}{2R_2}, \tag{3.45}$$

where $r$ is the radius of the planetary assembly. With this condition, the deposition rate is

$$D = \sqrt{\frac{m}{2\pi k \rho^2}} \frac{p_e}{\sqrt{T}} \frac{A}{4\pi r^2}. \tag{3.46}$$

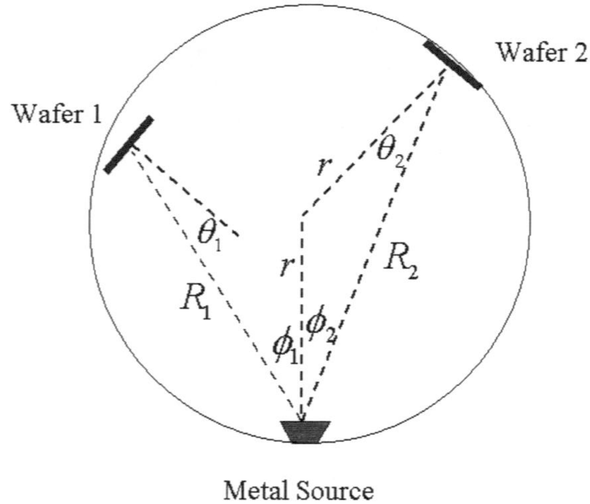


**Figure 13.** Cosine law of evaporation: Wafer 1 is in arbitrary location; Wafer 2 is on an imaginary sphere with radius $r$.

### 3.6 Chemical Vapor Deposition (CVD)

Several other deposition techniques are also commonly used. These include thermal growth of silicon dioxide on silicon at approximately 1000-1175 °C and chemical vapor deposition of dielectric materials. The chemical vapor deposition process is usually carried out in low-pressure vessels (tubes), where the temperature is maintained at 600-800 °C to achieve sufficient deposition rates. The chemical process can by pyrolysis, in which a gas molecule is decomposed on the surface, thus coating it. An example of such a process is the deposition of Si from silane ($SiH_4$) at temperatures above 800 °C. An alternative to the high-temperature CVD is the plasma-enhanced CVD, where the reagent molecules are heated by the gas plasma, thus forming reactive radicals at lower temperatures. Figure 14 shows a plasma-enhanced chemical vapor deposition (PECVD) chamber. Typical deposition temperatures in PECVD are 200-600 °C. A list of a few

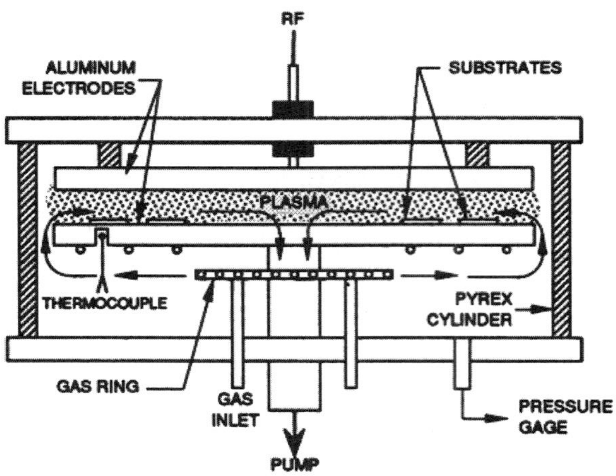


**Figure 14.** Radial flow plasma deposition chamber (reproduced with permission from Brodie and Muray (1992).)

commonly deposited materials and the reactive gases used are given in Table 3

**Table 3.** Commonly deposited inorganic films via PECVD from Brodie and Muray (1992).

| Film | Reagent Gases |
|---|---|
| Silicon Dioxide, $SiO_2$ | $SiH_4$ $+N_2O$ |
| Silicon Nitride, $Si_3N_4$ | $SiH_4+N_2$ |
| Amorphous Si, (a-Si) | $SiH_4$ |
| Silicon Carbide, SiC | $SiH_4+C_2H_4$ |

# 4 Surface Micro-Machining

Surface micro-machining refers to a fabrication process that generates mechanical structures on the surface of the substrate. Surface micro-machining, as defined today, originates from the first resonant gate field effect transistor (RGT) invented by Nathanson and Wickstrom (1965) of Westinghouse Research Laboratories. Since then, many other devices have been constructed including cantilevers, micro-bridges, springs, hinges, and other elastically suspended structures. In addition to the commonly used processes for the manufacturing of IC-s, electroplating and electroforming has also emerged as a viable process for fabricating surface micromachined structures. An example of such a process is the creation of an electroplated thermal actuator through a robust single-mask fabrication process (Enikov and Lazarov, 2003). The entire process flow is shown in Figure 15. The starting substrate is silicon, with 0.4-$\mu$m-thick thermally grown oxide. The cantilever beam (thermal actuator) is fabricated via electroplating of nickel in a mold of positive photoresist AZ4903. To release the beam, a so-called sacrificial layer is needed. For this purpose, a thin layer of 5nm Ti/100nm Cu was deposited via e-beam evaporation. This layer also serves as a seed layer for the electroplating step. The very thin titanium layer used under the copper provides adhesion to the $SiO_2$ through the formation of titanium oxide at the interface. Photoresist with a thickness in the range of 25-30 $\mu$m is spun and patterned in order to create an electroplating mold for the devices. Approximately 20 $\mu$m of nickel is then electroplated from a commercially available Ni bath (Microfab Ni 100, Enthone-OMI, Inc.). After removal of the photoresist, the actuators are released via wet etching of the Ti/Cu seed layer in an aqueous solution of ammonium persulfate. The lateral undercut of the narrower Ni structures releases them completely from the substrate while the wider features, such as bonding pads, remain anchored to the substrate. This process is analogous to the surface microcmachining of polysilicon structures, where a sacrificial oxide is used instead of the copper seed layer. An SEM micrograph of the completed device is shown in Figure 16. The device consists of a thermal actuator (right), connected to a moveable frame, whose position is detected by a column of interdigitated electrodes (left). The inset in the figure shows a magnified view of the capacitor comb-fingers.

As evident from the example above, surface micro-machining, which utilizes the standard sequence for pattern transfer used in IC manufacturing requires the patterned film to be deposited first, followed by a photolithography step and etching, as illustrated previously in Figure 5. In MEMS fabrication, however, an alternative technique known as "lift-off" patterning is also used. This technique requires deposition and patterning of the photoresist first, followed by deposition of the film to be patterned. The actual patterning occurs during the removal of the photoresist, as it lifts off the portions of the film not in direct contact with the substrate. The process is illustrated in Figure 17. This pattern transfer technique can be used with noble or inert metals such as Pt and W, for which there are no good chemical etchants. Further, the feature size is not affected by the lateral undercut occurring during etching, thus allowing printing of smaller features. The disadvantages of this technique is that if there are large continuous areas of photoresist, it may take too long to lift it off and the thickness of the patterned film is limited to about 1/4 to 1/3 of the thickness of the photoresist.

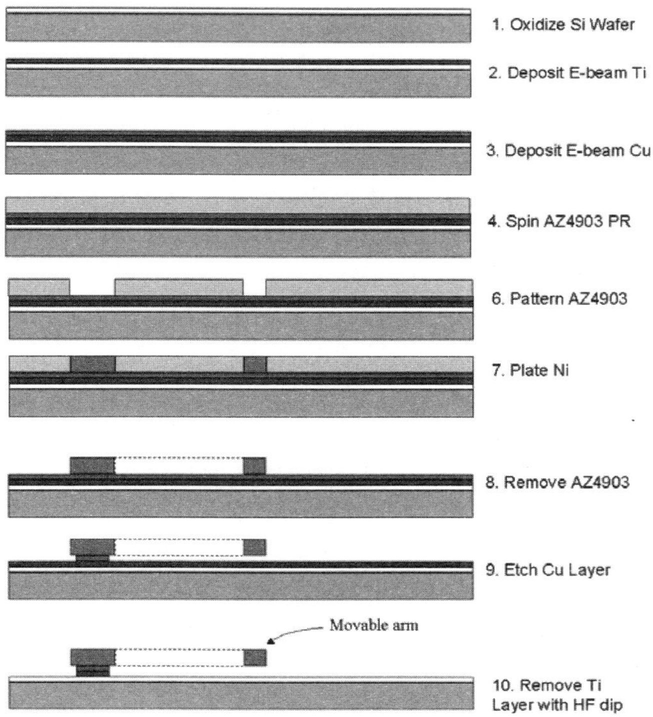


**Figure 15.** Fabrication sequence.

## 5 Bulk Micro-machining

Bulk micro-machining refers to a process of forming structures in the bulk of the substrate as opposed to surface structure formation. Subtractive pattern transfer (etching) is one of the most common processes used to delineate features in the substrate. The etching process can be a wet (liquid phase) or dry (gaseous phase) chemical etch or a physical etch (sputtering), where energetic ions are used to remove the material at a desired location. Several terms are commonly used to describe the etch process:

- *Bias* is the difference between the defined lateral dimension and undercut lateral dimension ($2D$ in Figure 18).
- *Tolerance* is the standard deviation of the bias over different locations on the substrate. It measures the uniformity of the etching over the substrate.
- *Etch rate* is equal to the etched depth per unit of time. The usual units are $\mu$m/min or $\mathring{A}$/sec.
- *Etch rate uniformity* is defined as the ratio between the maximum variation of etch rate and the average etch rate along the surface of the substrate.
- *Over etch* is the amount of additional etch time specified as a percentage of the nominal etch time needed to completely pattern the film.

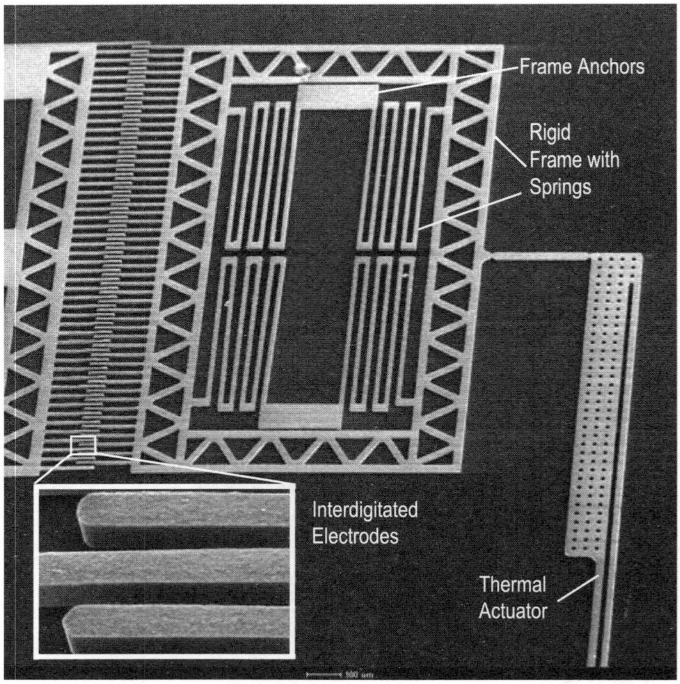


**Figure 16.** Surface micromachined Ni structures.

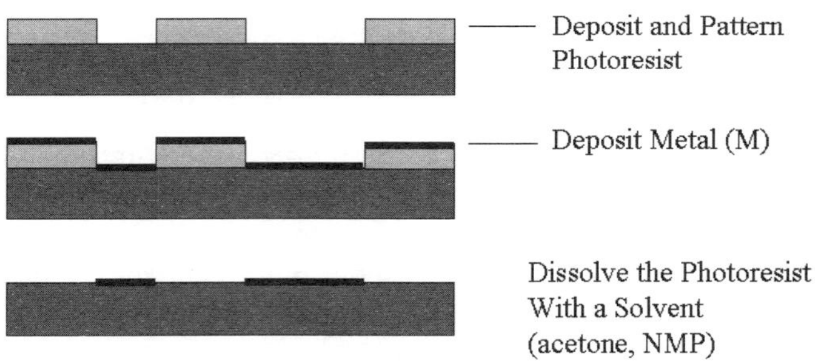


**Figure 17.** Lift-off process with positive resist

- *Selectivity* is usually defined as the ratio between the etch rate of the film and the etch rate of the masking material or the underlying substrate. Usually the selectivity is much greater than one.

Among all possible etching processes, two are particularly useful in the fabrication of MEMS: wet and dry etching of silicon. Two variants exist for each of these processes: isotropic etching and anisotropic etching. In isotropic etching, the etch rate is uniform in all directions while in anisotropic etching the etch proceeds much faster in some directions compared to others (see Fig. 18).

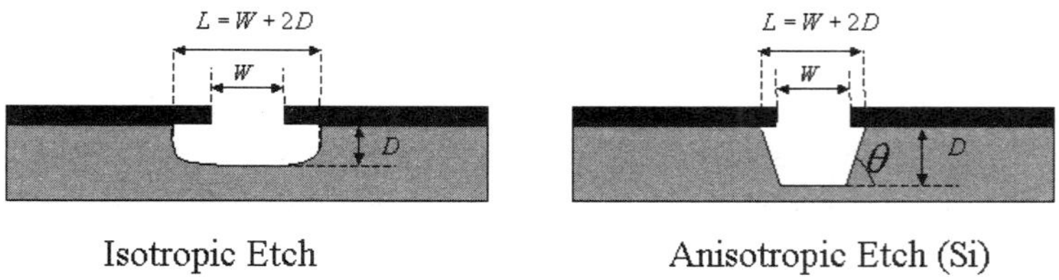


$$\theta = \cos^{-1}\frac{1}{\sqrt{3}} = 54.7°$$

**Figure 18.** Isotropic and anisotropic etching.

### 5.1 Isotropic Etching of Silicon

This etch is performed in a mixture of hydrofluoric and nitric acids in water. The overall reaction of this process is

$$\mathrm{Si} + \mathrm{HNO_3} + 6\mathrm{HF} \rightarrow \mathrm{H_2SiF_6} + \mathrm{H_2} + \mathrm{H_2O} + \mathrm{HNO_2}. \quad (5.1)$$

During this reaction, two concurring processes are underway: oxidation/reduction of Si by $HNO_3$ and dissolution of $SiO_2$ by HF. Thus, the etch rate is controlled by diffusion at high concentration of $HNO_3$ and by the formation of $SiO_2$ at high concentrations of HF.

### 5.2 Anisotropic Etching of Silicon

Wet anisotropic etching of silicon is performed in aqueous alkaline solutions of KOH or CsOH. The rate limiting step in this case is the thermal activation of electrons into the conduction band. Seidel et al. (1990) proposed a model for this process based on a multi-step reaction involving charge transfer electrons from the conduction band ($e_{cb}$)

$$\begin{aligned}
\mathrm{Si} : +2\mathrm{OH^-} &\rightarrow \mathrm{Si(OH)_2} + 2e_{cb}\\
\mathrm{Si(OH)_2} &\rightarrow \mathrm{Si(OH)_2^{2+}} + 2e_{cb}\\
\mathrm{Si(OH)_2^{2+}} + 2\mathrm{OH^-} &\rightarrow \mathrm{Si(OH)_4}\\
\mathrm{Si(OH)_4} &\rightarrow 2\mathrm{H^+} + \mathrm{Si(OH)_2O_2^{2-}}\\
2\mathrm{H^+} + 2\mathrm{OH^-} &\rightarrow 2\mathrm{H_2O}\\
4e_{cb} + 2\mathrm{H_2O} &\rightarrow 4\mathrm{OH^-} + 2\mathrm{H_2}
\end{aligned} \quad (5.2)$$

The overall etching reaction can be represented as

$$\mathrm{Si} : +2\mathrm{H_2O} + 2\mathrm{OH^-} \rightarrow \mathrm{Si(OH)_2^{2+}O_2^{2-}} + 2\mathrm{H_2}, \quad (5.3)$$

where the reaction rate $R$ depends on the local concentration of reactants and products and the activation energy $E_a$

$$R = R_0 \frac{[\mathrm{H_2O}]^2[\mathrm{OH^-}]^2}{[\mathrm{Si(OH)_2^{2+}}][\mathrm{H_2}]^2}, \qquad R_0 = Ae^{-E_0/kT}. \tag{5.4}$$

In Seidel et. al.'s model, the activation energy $E_a$ is different for different crystal planes, resulting in 200-300 times faster etch rates for (100) planes compared to (111) planes of Si. This results in a formation of crystallographically-defined planes as illustrated in Fig. 3(d).

### 5.3 Deep Reactive Ion Etching (DRIE) of Silicon

Reactive ion etching of Si is one of the common semiconductor processing steps used for pattern transfer or film removal. The majority of conventional plasma-assisted etching processes are isotropic, which limits their applicability to shallow pattern transfers only. The demand for high-aspect-ratio structures used in electrostatic, fluidic, and other MEMS devices resulted in the development of a new processing tool for deep reactive ion etching (DRIE). This powerful bulk micro-machining technique allows the etching of structures with aspect ratios (depth/width) of up to 20, as shown in Figure 19. Such structures have become the fundamental building blocks of many MEMS devices, ranging from capacitive force sensors, rate gyros, and comb drives for optical MEMS to bioMEMS in which channel networks in silicon are used directly or as a molding template for the fabrication of polymeric MEMS. The anisotropy of the DRIE is achieved through a

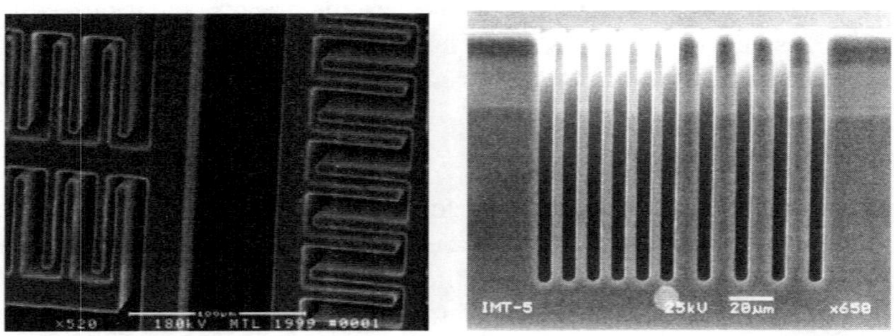


**Figure 19.** High aspect ratio Si structures: comb capacitors (left); test trenches (right).

fluorine-based chemistry process invented by Robert Bosch GmbH, which, instead of simultaneous passivation and etching, uses the concept of sequentially alternating two isolated mechanisms–etch and deposition. During Phase 1 (deposition), a passivation layer ($C_xF_y$) is deposited on the sidewall and the base of the trench by ionization and dissociation of octafluorocylcobutane ($C_4F_8$) through the following reactions:

Plasma reaction: $C_4F_8 + e^- \rightarrow CF_x^+ + CF_x^{\cdot} + F^{\cdot} + e^-$

Passivation layer deposition: $CF_x^{\cdot} \rightarrow nCF_2$(fluorocarbon polymer).

Due to the directional action of ions, the passivation layer on the bottom of the trench is etched away by ion-assisted fluorine radicals:

$$nCF_2 + F^{\cdot} \rightarrow CF_x^{\cdot} \rightarrow CF_2(\text{gas}).$$

During Phase 2 (etch phase), sulphur hexafluoride ($SF_6$) is used to spontaneously etch silicon isotropically by initially dissociating the relatively inert gas molecules into atomic fluorine radicals through the plasma reaction

$$SF_6 + e^- \rightarrow S_xF_y^+ + S_xF_y^{\cdot} + e^-.$$

The freed fluorine radicals are then used to etch the silicon:

$$\text{Si} + F^{\cdot} \rightarrow \text{Si}F_x(\text{gas}).$$

The trench propagates by sequentially alternating between Phase 1 and Phase 2, which enables the etching to proceed vertically by restricting the lateral etching. This switching process mechanism, however, leads to scallops on the sidewalls, noticeable for example on Figure 20 (right). Adjusting the process parameters of the two phases can minimize this effect.

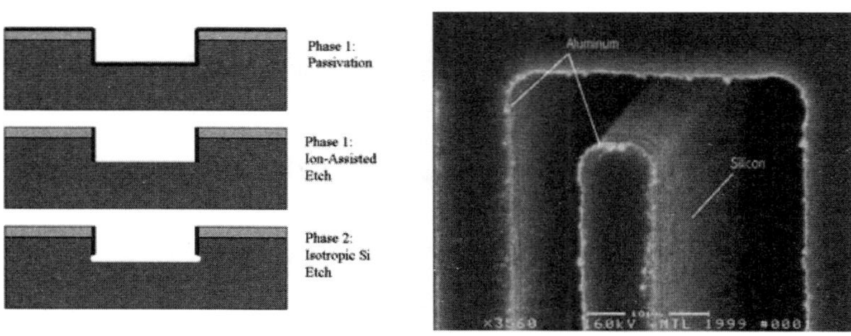


**Figure 20.** Bosch process: Phase 1 and Phase 2 (left); resulting scallops (right).

In comparison to other bulk micro-machining techniques, the DRIE process can produce any two-dimensional geometry and is not restricted by the crystalline orientation of the substrate. Similarly to wet etching, this process also terminates on silicon dioxide and silicon nitride, allowing these materials to be used as masks or etch limiting layers. This is especially important in through-the-wafer etching of micro-nozzles and channels.

## 6 Other Micro-fabrication Techniques

In addition to the traditional processes borrowed from the manufacturing of integrated circuits, many other fabrication techniques have emerged to fulfill the needs of heterogeneous MEMS devices. While a complete review of these is beyond the scope of this lecture, the most important techniques are briefly described in this section. These include high-aspect-ratio structures fabricated via LIGA, techniques for depositing biological materials through stamping (soft-lithography), and bonding methods used to create glass-metal or glass-silicon seals adopted in MEMS to create a variety of micro-fluidic systems.

### 6.1 LIGA

The acronym LIGA was coined from the German words for lithography (lithographie), electrodeposition (galvanoformung), and molding (abformtechnik). The process dates back to 1970s when IBM first used electroplating in X-ray patterned masks to form magnetic recording heads(Romankiw et al., 1970). Later, the molding process was added to by researchers at the Kalrsruhe Nuclear Research Center (KfK), who were interested in producing micron-sized nozzles for separation of uranium 235 (Becker et al., 1982). Decades later, with the emergence of MEMS, interest in this technology resurged. Today, networks of laboratories capable of generating synchrotron radiation are offering manufacturing services based on LIGA.

The key feature of LIGA is its ability to produce high-aspect-ratio structures. For example, vertical pillars and trenches a few microns wide and several hundreds microns tall are routinely produced. This is a result of the short wavelength of the X-ray radiation (2-10 $\mathring{A}$), low scattering and diffraction, and large penetration depth in polymeric materials. In principle, the process is somewhat similar to conventional lithography in that a two-dimensional mask is used to define a pattern in a poly(methylmethacrylate) (PMMA) layer (resist). The patterned structure is then filled with metal (nickel in most cases), after which the polymer is dissolved. The resulting Ni features can be used directly or, as in the original LIGA process, as a mold for the fabrication of replicas of the original mask pattern through hot embossing or casting. The X-rays used in LIGA are usually from synchrotron source, which is $10^4$ times more powerful than a rotating disk anode source. These X-rays are emitted by high-energy relativistic electrons in a storage ring accelerated in the normal direction by a magnetic field, $B$. The radiated power from an electron current, $i$, with energy $E$ (in GeV), is given by (Brodie and Muray, 1992)

$$P(kW) = \frac{88.47E^4 i}{\rho}. \tag{6.1}$$

Typical values for the electron energies are $E = 3.5$ GeV, $\rho = 12.7$ m, and $i$=0.1 A (data is for the SPEAR facility at Stanford), resulting in 105 kW of radiated power. The synchrotron radiation is emitted in a cone with an angle $\alpha$ (see Fig. 21)

$$\alpha = \frac{0.5}{E(\mathrm{GeV})}(\mathrm{mrad}). \tag{6.2}$$

The emitted radiation has a broad spectrum, from microwave down to the X-ray wavelengths. Since the penetration depth increases with the photon energy, the top portions of the PMMA resist will get a higher energy dose, which can result in resist damage. To avoid this, a pre-absorber is placed in front of the x-ray mask to filter out low-energy photons. The mask usually consists of a Si or Be membrane, carrying gold or tungsten features that absorb the incident radiation. Often the absorbing material is much thicker than the membrane itself, which makes the fabrication of such masks very costly ($15-30K).

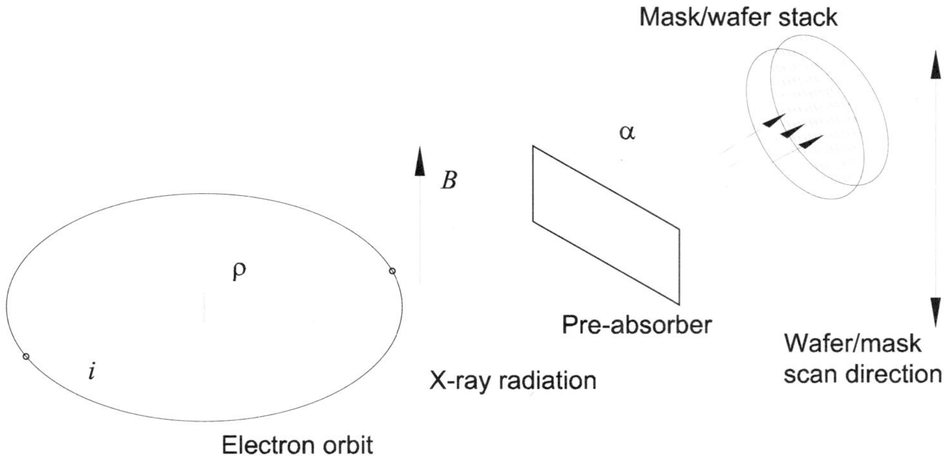


**Figure 21.** LIGA

### 6.2 Soft-lithography

Soft lithography has gained increased popularity in the last five years due to its ability to pattenr proteins and other biological materials with sub-micron resolution. This technique involves the fabrication of a poly(dimethylsiloxane) (PDMS) stamp through a photolithographic technique. The PDMS is moderately stiff, with a Young's modulus of 1 MPa. It is available commercially under the tradename Sylgard 184 and is nontoxic and intrinsically very hydrophobic. Through plasma oxidation, the surface of PDMS can be converted to a high surface-energy form, negatively charged, which is hydrophilic. In this form, it can be used in capillary electrophoresis applications or as a simple stamp used to transfer patterns, similar to the conventional ink-printing technology. The process flow for stamp formation as described by Whitesides et al. (2001) is shown in Figure 22. Stamp fabrication begins with a conventional photolithographic step defining a photoresist "master" on a silicon wafer. Then, the cured photoresist is exposed to vapors of $CF_3(CF_2)_6(CH_2)_2SiCl_3$ for 24 hours to reduce its tendency to adhere to the stamp. A Sylgard 184 elastomer is cast over the photoresist and cured for 2 hours at 60 °C. The cured PDMS stamp is peeled off and inked with an ethanol solution of alkanethiols. The ethanol is driven out by drying the master in a nitrogen atmosphere to prevent oxidation of the thiols. Finally, the master is pressed against the gold film for a period of 30 seconds to 5 minutes, resulting in a 2-nm-thick print of a SAM. Washing with a different alkanethiol can be used to coat the unexposed areas with a complementary (hydrophilic)

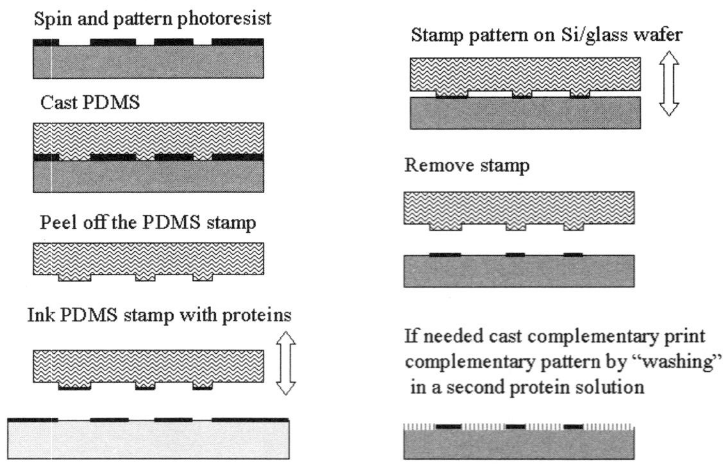


**Figure 22.** Soft ligthography process (after Whitesides et al. (2001)).

SAM. Following this process Whitesides et al. (2001) attained a minimum line width of 100 nm.

### 6.3 Anodic Bonding

Anodic bonding is a common method of bonding glass to conductors such as silicon or metal. The glass and conductor are brought into contact at elevated temperatures and an electric field is applied with the cathode on the glass and the anode on the conductor (see Fig. 23). The mobile positive ions, typically 4 mol% $Na_2O$ in borosilicate glass (Pyrex), diffuse toward the cathode, leaving behind the negatively charged oxygen ions in close proximity to the glass/conductor interface ( see Fig. 24). A positive image charge develops on the conductor surface, and the electrostatic attraction eventually pulls the glass and conductor together with a pressure high enough to initiate a surface reaction and form a chemical bond. An anodic bond is possible only under certain conditions (Ko et al., 1985). The surfaces need to be sufficiently smooth and the thermal expansion coefficients of the two materials should be matched within 2 ppm, since a larger mismatch will result in bond failure upon cooling of the stack. Finally, in order to develop a space charge near the surface, the conductor should be a "blocking electrode," i.e., should not introduce positive charge carriers into the glass. Silicon and some inert metals such as Kovar alloy (for composition, see below) are blocking electrodes and thus meet these requirements.

Unfortunately, anodic bonding is performed under the harsh conditions of high temperature and pressure, and an electric field, leading to a complicated micro-structural

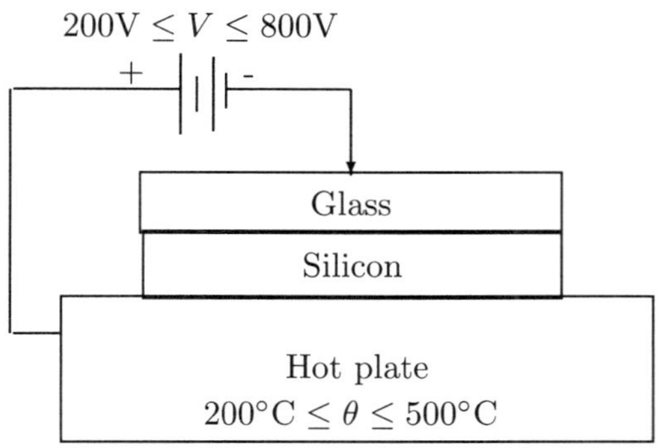


**Figure 23.** Bonding Setup

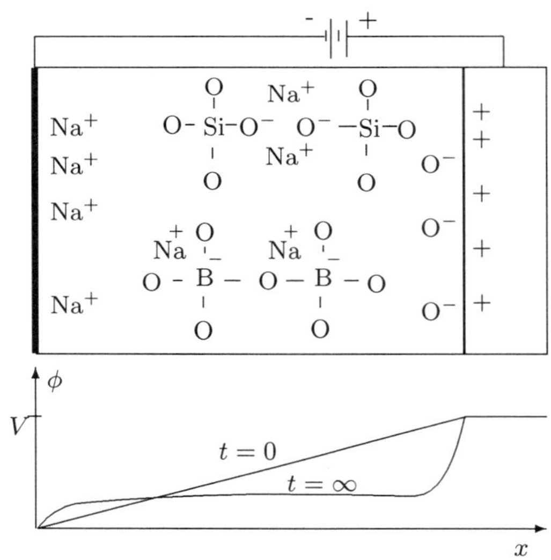


**Figure 24.** Ionic Process in Pyrex

evolution including creep, ionic diffusion, residual stresses, warping, and spatial instabilities such as electrical breakdown and symmetry breaking of the bond front. In particular, the extreme temperature conditions limit the choice of materials that can be used in MEMS. For example, a micro-pump with a polyimide diaphragm was sealed via anodic bonding under an inert atmosphere in order to prevent oxidative damage (Lin et al., 1996; Hesketh et al., 1996). The harsh conditions of the bonding step can also affect the device characteristics after bonding. Since the typical bonding temperature is 400 °C, the locked-in residual stresses can offset the characteristics of any piezo-resistive strain gauges, or the gap of a capacitive accelerometer. Thus a good understanding of the conditions of anodic bonding is imperative for the achievement of a high-yield bonding.

The first model of anodic bonding was introduced by the inventors of this technique (Wallis and Pomerantz, 1969). The inventors report that it is possible to make seals between ceramics such as soda lime #0080, potash soda lead #0120, aluminosilicate #1720 fussed alumina, optical fibers and a matching metal such as Tantalum, titanium, Kovar, Nitromet 44, and also silicon, germanium, and gallium arsenide. Further, thin films (less than 2 mil) of aluminum, nickel-chromium, iron, boron, silicon monoxide, or thermally grown silicon dioxide can also be bonded to these glasses. In some cases, glass-to-glass seals are also possible. The initial gap after contact is on the order of a micron. Further, it was observed that the attractive force appears only when the glass is at negative potential with respect to the silicon. An electron micrograph of the bonded interface between Pyrex and silicon showed no transition layer between the two regions i.e., the bond is not due to interdiffusion. Wallis and Pomerantz proposed a simple two degree of freedom, lumped parameter model for the electric variables (shown in Figure 25). Based on this electric circuit, they estimated the electrostatic pressure between

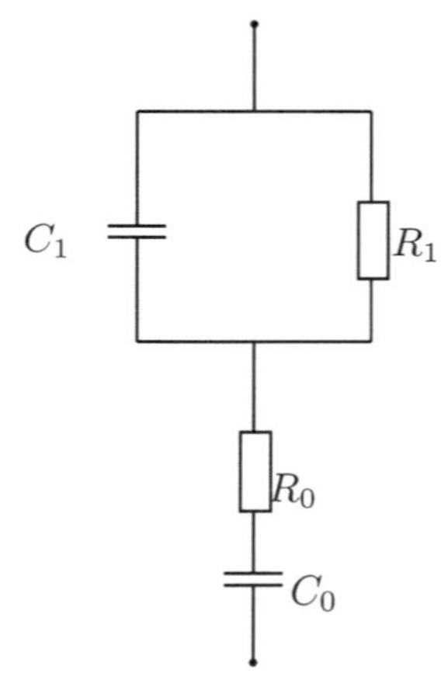


**Figure 25.** Two degree of freedom model.

Pyrex and silicon at 300 °C of 800V applied potential at 350 Psi, which corresponds to an electric field of $3 \times 10^6$V/cm. This is close to the electric breakdown of glass. The required time to produce a seal after an intimate contact at different temperatures has been interpreted as a sign of the chemical nature of the bond. At 550 °C a Pyrex-silicon seal forms instantaneously, where at 180 °C it takes many hours. The strength of the bond was established as 1500-3000 Psi. Further insights into the nature of the bonding process were provided by Borom (1973). He studied the bonding of Fe substrates to glass with and without an applied electric field. Comparing the two cases, Borom discovered that the $Fe^{2+}$ transport in glass is both concentration and field driven. No bond was observed, however, which was explained by the lack of FeO at the interface. On the other hand, the penetration depth of Fe in both cases was identical and the ratio of the concentrations constant. In both cases, Borom found the diffusion coefficients of Fe to be equal. Hence, he concluded that Fe diffuses in an electrically neutral form. The lack of a bond is explained by the slow Fe oxidation at the interface and by the faster dissolution of FeO into the bulk of the glass, confirming the need for a blocking electrode that does not introduce carriers into the glass.

A more detailed study by Albaugh and Rasmussen (1992) showed experimentally that the bonding process is through contact area enlargement rather than by development of a uniformly increasing interfacial bond. Another important conclusion of this work is that the kinetically limiting process is the sodium mobility in the bulk region. The charge transport of other ions (like oxygen) within the depletion region is rapid once a high enough electric field is established there. This explains why, at lower temperatures, there is a latency time of about 5 to 15 seconds before bonding occurs–time necessary to build the electric field. From polarization studies, Albaugh and Rasmussen concluded that oxygen is delivered to the anode at all times, but only when intimate contact is achieved is this oxygen utilized to form an oxide (bond). During 80% of bond formation approximately 2-5 $mC/cm^2$ of charge is transferred. An equivalent amount of oxygen (2 $mC/cm^2$) results in approximately 1.4 nm of silicon oxide growth (a few monolayers). These estimates support the observation of Wallis and Pomerantz (1969) that no transition (inter-diffused) layer is formed. Albaugh and Rasmussen further estimated the activation energy for bond formation to be $0.66\pm0.1$eV (sodium diffusion in silica has an activation energy of 0.7 to 1.5eV, the viscous flow has activation energy of a few electron-volts, and the energy of network dissociation is about 17eV). Thus, the rate-limiting step should be the sodium conduction and not the dissociation or viscous deformation.

Studies on the nature of the sodium depletion layer in the glass showed that it is significantly less dense since alkali ions and non-bonded oxygen diffused out (Carlson, 1974). This layer sinters (densifies) at a rate that is dependent on the glass composition, temperature, and applied field. Further, the author concludes that an electrochemical oxidation takes place since a fixed amount of charge is required for bond formation.

The actual mechanism of bond formation is still under debate. The electrochemical oxidation theory (charge transfer being a measure for the bond progress) was disputed by Morsy et al. (1996). Their group bonded glass #0700 wafers from Matsunami Co. Ltd. ($SiO_2$70 %;$Al_2O_3$4%; $B_2O_3$9%; $Na_2O$2–4%; $K_2O$2–4%; $Li_2O$2–4%; CaO 2–4%; BaO 2–4%) to a Kovar alloy (C 0.003%; Co 16.2%; Fe 53.65%; Mn 0.45%; Ni 29.6%; Si 0.1%). At 563 K, a 180-second delay was observed in the initiation of bond spreading. At 613 K, however, there was no delay. Defining a complete bond as 80% of the bonded area, the authors calculated activation energies of 460 kJ/mol for 670-680 K and 65 kJ/mol below this temperature. The large variation of the activation energies in the two temperature regions points toward two different rate-limiting mechanisms. Charge transfer alone does not measure the progress of the bond formation (the process is not purely electrochemical oxidation). The activation energy at high temperature is close to the energy for viscous flow (430 kJ/mol) where the low-temperature one is close to the ionic conductivity of the glass (65 kJ/mol), i.e., the viscous flow is the rate limiting process at the high-temperature range, but it is bulk conduction at low temperatures.

Despite the uncertain nature of the underlying mechanism, anodic bonding is widely used as a relatively low-temperature sealing process. The emergence of three-dimensional MEMS Structures, as well as planar electrophoretic channels, sparked a renewed interest in anodic bonding applied to stacks of glass and silicon. Tri-layer glass-silicon-glass bonds have been demonstrated by Hartz (1992) using voltage reversal (5 min in one direction and 5 min in the opposite direction). Harz applied 40 V at 520 °C to fabricate three passive check valves. The bonding of the opposite structure silicon-Pyrex-silicon, was also demonstrated by Feinerman et al. (1992) and Despont et al. (1995, 1996), who used it in the fabrication of a miniaturized electron column. Finally, Si-to-Si bonding with a 1-5 $\mu m$ sputtered borosilicate glass layer is also possible (Hanneborg et al., 1991; Spiering et al., 1995). The bond is performed at 400 °C and 200 V for 10 min. A bond strength of 2-3 MPa was achieved with a sputtered glass thickness of 2 $\mu m$. The use of such thin glass films allows bonds with even lower applied voltages. For example, 15 V was needed to start the bonding (Spiering et al., 1995) and even 1V was sufficient for thinner films (100nm).

## 7 Summary

A concise review of the main fabrication techniques for micro-systems has been presented following the material presented in the actual lectures during the CISM short course on Micro-Mechanical Systems Design (June 28 – July 2, 2004). While I have made every effort to concisely review the major processes used in MEMS fabrication, this review is not exhaustive. The interested reader is referred to the extensive review on MEMS fabrication documented by Madou (1997), as well as to the excellent textbooks on semiconductor physics by Sze (1981) and on integrated circuit manufacturing by Runyan and Bean (1990); Brodie and Muray (1992), and Wolf and Taubert (2000), which contain extensive treatments of this subject.

## Acknowledgement

This material is based upon the work supported by the National Science Foundation under Grants No. EEC-0407369, DMI-0134585, and DMI-0303868. Any opinions, findings, and conclusions or recommendations expressed in this material are those of the author(s) and do not necessarily reflect the views of the National Science Foundation.

## Bibliography

Albaugh, K. and Rasmussen, D.H. (1992) Rate processes durin anodic bonding. *Journal of the American Ceramic Society*, 75(10), pp. 2644–48.

Becker, E., Ehrfeld, W., znc H. Betz, D.M., Heuberger, A., Pongratz, S., Glashauser, W., Michel, H., and Siemens, V. (1982) Production of separation nozzle systems for uranium enrichment by a combination of x-ray lithography and galvanoplastics. *Naturwissenschaften*, 69, pp. 520–523.

Borom, M. (1973) Electron-microprobe study of field-assisted bonding of glasses to metals. *Journal of the American Ceramic Society*, 56, pp. 254–257.

Brodie, I. and Muray, J.J. (1992) *The Physics of Micro/Nano-Fabrication.* Plenum Press, New York and London.

Carlson, D.E. (1974) Ion depletion of glass at a blocking anode: I and ii, theory and experimental results for alkali silicate glasses. *Journal of the American Ceramic Society*, 57(7), pp. 291–300.

Caughey, D. and Thomas, R. (1967) Carrier mobilities in silicon empirically related to doping and field. *IEEE Proc.*, 55, pp. 2192–2193.

Deal, B.E. and Grove, A.S. (1965) General relationship for the thermal oxidation of silicon. *J. Appl. Phys.*, 36, pp. 3770–3778.

Despont, M., Gross, H., Arrouy, F., and Staufer, C.S.U. (1996) Fabrication of a silicon-pyrex-silicon stack by a.c. anodic bonding. *Sensors and Actuators A*, 55, pp. 219–224.

Despont, M., Staufer, U., Germann, R., and Vettiger, P. (1995) Microfabrication of lenses for a miniaturized electron column. *Microelectonics Engineering*, 27, pp. 467–470.

Enikov, E. and Lazarov, K. (2003) PCB-integrated metallic thermal micro-actuators. *Sens. Actuators A, Phys.*, 105(1), pp. 76–82.

Feinerman, A.D., Crewe, D.A., Perng, D.C., Shoaf, S.E., and Crewe, A.V. (1992) Sub-centimeter micromachined electron microscope. *Journal of Vacuum Science and Technology A*, 10(4), pp. 611–616.

Ghez, R. (2001) *Diffusion phenomena: cases and studies.* Kluwer Academic/Plenum Press, New York.

Hanneborg, A., martin Nese, and Ohlckers, P. (1991) Silicon to silicon anodic bonding with borosilicate glass layer. *Journal of Micromechanics and Microengineering*, 1, pp. 139–144.

Hartz, M. (1992) Anodic bonding for the third dimension. *Journal of Micromechanics and Microengineering*, 2, pp. 161–163.

Hesketh, P.J., Lin, Y., Boyd, J.G., Zsivanovic, S., Cunnen, J., Ming, Y., Stetter, J., Lunte, S.M., and Wilson, G.S. (1996) Biosensors and microfluidic systems. In *International Symposium on Aerospace Japan.*

Howe, R.T. (1980) Resonant microsensors. In *Technical Digest Transducers '87, 4th Int. Conf. on Solid-State Sensors and Actuators, Tokyo, Japan.* pp. 834–849.

Howe, R. and Muller, U. (1986) Resonant microbridge vapor sensor. *IEEE trans. Electon Devices*, ED-33, pp. 499–506.

Jansen, H., de Boer, M., Wensink, H., Kloeck, B., and Elwenspoek, M. (2001) The black silicon method. VIII. a study of the performance of etching silicon using $SF_6/O_2$-based chemistry with cryogenical wafer cooling and a high density icp source. *Microelectronics Journal*, 32(9), pp. 769–777.

Keller, C. and Howe, R. (1995) Nickel-filled HEXSIL thermally actuated tweezers. In *Transducers '95: Eighth International Conference on Solid-State Sensors and Actuators*, vol. 2. pp. 376–379.

Kilby, J.S. (1964) Miniaturized electronic circuit. *US Patent #3,138,743.*

Ko, W.H., Suminto, J.T., and Yeh, G.J. (1985) Bonding techniques for microsensors. *Micromachining and Micropackaging of Transducers*, -, pp. –.

Kurtz, A.D. and Goodman, S.J. (1974) Apparatus and method for interconnecting leads in a high temperature pressure transducer. *US Patent #3,800,264.*

Lin, Y.C., Hesketh, P.J., Boyd, J.G., Lunte, S.M., and Wilson, G.S. (1996) Characteristics of a polyimide microvalve. In *Solid-State Sensor and Actuator Workshop*. Hilton Head, South Carolina, June 2-6, pp. 113–116.

Madou, M. (1997) *Fundamentals of Microfabrication*. CRC Press, Boca Raton, FL.

Menz, W. (1992) Microactuators in liga technique. *Int. J. App. Alectromagn. Mater.*, 2(4), pp. 281–284.

Meyerhofer, D. (1978) Characteristics of resist films produced by spinning. *Journal of Applied Physics*, 49, pp. 3993–3997.

Morin, F. and Maita, J. (1954) Electrical properties of silicon containing arsenic and boron. *Phys. Rev.*, 96, pp. 28–35.

Morsy, M.A., Ikeuchi, K., Ushio, M., and Abe, H. (1996) Mechanism of enlargement of intimately contacted area in anodic bonding of kovar alloy to borosilicate glass. *Materials Transactions, JIM*, 37(9), pp. 1511–1517.

Nathanson, H. and Wickstrom, R.A. (1965) A resosnant gate transistor with high-q bandpass properties. *Applied Physics Letters*, 7(4), pp. 84–86.

Nicollian, E.H. and Brews, J.R. (1982) *MOS (Metal Oxide Semiconductor) Physics and Technology*. John Wiley & Sons, New York.

Peterson, K. (1982) Silicon as a mechanical material. In *Proc. SPIE*, vol. 70. pp. 420–457.

Romankiw, L.T., Croll, I.M., and Hatzakis, M. (1970) Batch-fabricated thin-film magnetic recording heads. *IEEE Trans. Magn.*, MAG-6, pp. 597–601.

Runyan, W.R. and Bean, K.E. (1990) *Semiconductor integrated circuit processing technology*. Addison-Wesley Publishing Company, Inc., New York.

Sadler, D., Gupta, S., and Ahn, C. (2001) Transformers, inductors, power and magnetic device control–micromachined spiral inductors using uv-liga techniques. *IEEE Trans. Magn.*, 37(4), pp. 2897–2899.

Seidel, H., Hueberger, A., and Baumgartel, H. (1990) Anisotropic etching of crystalline silicon in alkaline-solutions. 1. orientation dependence and behavior of passivation layers. *Journal of the electrochemical society*, 137, pp. 3612–3626.

Spiering, V., Berenschot, J., and Fluitman, J. (1995) Sacrificial wafer bonding for planarization after very deep etching. *Journal of Microelectromechanical Systems*, 4(3), pp. 151–157.

Sze, S.M. (1981) *Physics of semiconductor devices*. John Wiley & Sons, New York.

Waggener, H., Kragness, R.C., and Tyler, A. (1967) Anisotropic etching for forming isolation slots. *Electronics*, 40, p. 274.

Wallis, G. and Pomerantz, D. (1969) Field assisted glass-metal sealing. *Journal of Applied Physics*, 10(10), pp. 3946–3949.

Whitesides, G., Ostuni, E., Takayama, S., Jiang, X., and Ingber, D.E. (2001) Soft lithography in biology and biochemistry. *Annual Reviews of Biomedical Engineering*, 3, pp. 353–373.

Wolf, S. and Taubert, R. (2000) *Silicon Processing for the VLSI era*. Lattice Press, Sunset Beach, CA.

# Microstructures Under Electrostatic Loads: Discrete System Modelling

Francesco De Bona

Dipartimento di Ingegneria Elettrica Gestionale e Meccanica, University of Udine, Udine, Italy

**Abstract.** Due to a scaling effect, electrostatic forces, usually negligible at macro-scale, become relevant at micro-scale. It follows that electrostatic actuation is used very often for microsystems. The evaluation of the mechanical behavior of microstructures under electrostatic forces requires a new approach based on a so-called coupled field analysis; in fact, due to electrostatic forces structure exhibits a deformation that generally influences the electrical field and therefore again the electrostatic forces themselves. The case of a single degree of freedom electromechanical system was first considered; as generally the case of continuum structure is developed by referring to a FEM discretisation, the more general case of a multi degrees of freedom system was then considered.

## 1 Scale Effects in Microsystem Design

Very often the mechanical design of microsystems can be performed using the same tools already developed at macro level. In other cases new mechanical design approaches are required. This is mainly due to scale effects. There are generally two types of scaling laws that are applicable to the design of microsystems (see, e.g., Hsu Tai-Ran, 2002):

- scaling laws strictly dependent on the size of physical objects (es: scaling of geometry, electrostatic and electromagnetic forces)
- scaling laws involving the scaling of the phenomena affecting the behaviour of microsystems.

A typical example of the second kind is the case of fatigue design. Silicon, one of the most common material used for microfabrication, should not be sensitive to fatigue in air at room temperature, however, thin films (2 to 20 μm thick) of silicon are known to fail prematurely under cyclic fatigue loading in room-temperature air. The mechanism is not clear, in fact silicon does not display the room temperature plasticity or extrinsic toughening mechanisms necessary to cause fatigue in either ductile (e.g. metals) or brittle (e.g. ceramics) materials. Very recently Muhlstein et al. (2002) stated that cycling fatigue of thin layers of silicon seems related to a conceptually new and different mechanism termed "reaction –layer" fatigue.

Scaling in geometry can be easily understood considering that as volume $V$ scales with $l^3$, being $l$ a characteristic length reference, and surface $S$ with $l^2$, it follows that $S/V$ scales with $l^{-1}$. It can be thus possible to conclude that a reduction of size of 10 times will mean a 1000 times reduction in volume, but only a 100 time reduction in surface area. Thus motivated, for instance, the great effort applied for the development of MAV (Micro Air Vehicle) mainly for military applications (Dornheim, 1998).

Probably the main relevant type of scaling law strictly dependent on the size of physical objects is that related to the actuation forces. A thoroughly treatise of this subject was firstly developed by Trimmer (1989); in particular a matrix form was proposed (Trimmer's force scaling vector) to made clear how the different forces scale into the micro-domain. Among the different kinds of actuating principle which benefit of scale effect, the most widely used is electrostatic actuation. As it is well known, if the case of a simple parallel capacitor is considered, the electrostatic force in the $z$ direction normal to the plate surface is:

$$F_z = -\frac{\varepsilon_r \varepsilon_0 w l}{2 g_0^2} V_0^2 \qquad (1\text{-}1)$$

where:

$V_0$= applied voltage
$\varepsilon_0$= permittivity of vacuum
$\varepsilon_r$= relative permittivity of the dielectric
$l$, $w$ =plate dimensions in $x$ and $y$ directions
$g_0$=gap between the two parallel plates.

If one plate of the capacitor is moved parallel to the other a force tending to realign the plates occurs respectively in the $x$ or $y$ directions:

$$F_x = -\frac{\varepsilon_r \varepsilon_0 l}{2 g_0} V_0^2 \qquad F_y = -\frac{\varepsilon_r \varepsilon_0 w}{2 g_0} V_0^2 \qquad (1\text{-}2)$$

These forces can be used to obtain different types of actuators. Electrostatic forces in $x$ and $y$ directions are used to obtain tangential force actuators, that are called comb drives, when, to increase the actuating force, several elements are arranged in parallel. If the actuating force is normal to the capacitor surface, the actuator is called normal force actuator. According to the previously reported expressions, electrostatic force scales with $l^0$ (see, e.g., Fukuda, 1998); considering that other kinds of force (electromagnetic, piezoelectric, inertial, etc.) scale with a positive power of $l$, it follows that they strongly reduce their intensity as the characteristic dimension of the device reduces. On the other hand in Madou (2002) it is observed that a scale effect has to be considered also for voltage. In fact it is well known that the maximum value of voltage that can be applied to a parallel plate actuator, called breakdown voltage, is the voltage applied between a pair of electrodes for which the insulating properties of the dielectric material are destroyed, leading passage of a high current, which is accompanied by the emission of heat, light and sound, that is a spark discharge. The breakdown characteristic of an electrode gap is function of the product of gas pressure and gap distance (Paschen's law ) as shown in Fig.1.

According to Paschen's law the breakdown voltage follows different scaling law: in the case of great dimensions, it decreases with the gap, reaching a minimum, after which a further decrease in the gap dimension produces a proportional increase of the value of the breakdown voltage; this seems related to the fact that the electrode spacing is too small for ionization to occur; when the pressure is too low or the distance too small, most electrons reach the anode without colliding with gas molecules (Madou, 2002). From the design point of view it is particularly important to know the value of the minimum breakdown voltage. Table I reports this values for various gases.

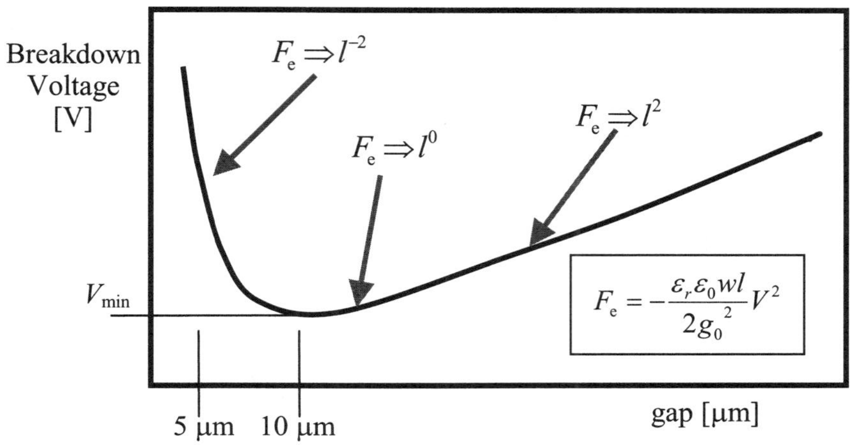


**Figure 1.** Paschen's curve and scaling law of electrostatic actuation.

It is therefore possible to conclude that, depending from the gap size, different scaling law of the electrostatic forces are possible; as shown in Figure 1, at lower values of the gap $F_e$ scales with $l^{-2}$, at higher values of gap it scales with $l^2$, whereas in proximity of the minimum electrostatic forces are constant with respect to the device dimensions.

**Table 1.** Minimum breakdown voltage for different gases.

| Gas | $V_{min}$ [Volt] |
|---|---|
| air | 327 |
| Ar | 137 |
| $H_2$ | 273 |
| He | 156 |
| $CO_2$ | 420 |
| $N_2$ | 251 |
| $O_2$ | 450 |

Apart from scaling considerations, electrostatic actuation seems to be more suitable for microsystems in comparison to electromagnetic actuation also because electrostatic motors are simpler and more compatible with IC fabrication. Moreover the electrostatic actuator is driven by voltage, and voltage switching is easier and faster than current switching (as in electromagnetic actuators) at microscale. An implicit confirmation of the previous assessments follows from the wide numbers of microdevices that are based on electrostatic actuation. That is the case of different kind of rotary motors (Madou, 2002). A solution based on a electrostatic actuator is used in the most popular commercial applications of microsystems, that is the self test unit of the ADXL accelerometers from Analog Devices and micromirror actuating unit for the Texas Instruments projection display (Senturia, 2002). Electrostatic actuation is widely used also in the case of RF MEMs, to actuate switch, filters, etc. (Nguyen, 2000).

From the engineering design point of view it is therefore important to handle suitable models of such devices; this aspect is relatively new, as electrostatic forces are negligible at macro scale and therefore the usual methods developed to deals with macro scale device can not by applied at the microlevel. In the following a possible approach will be proposed. At first the case of a single degree of freedom system, than the case of discrete systems with several degrees of freedom will be considered.

## 2 "Single" Degree of Freedom System: Static Analysis

Figure 2 shows a "single" [1] degree of freedom system.

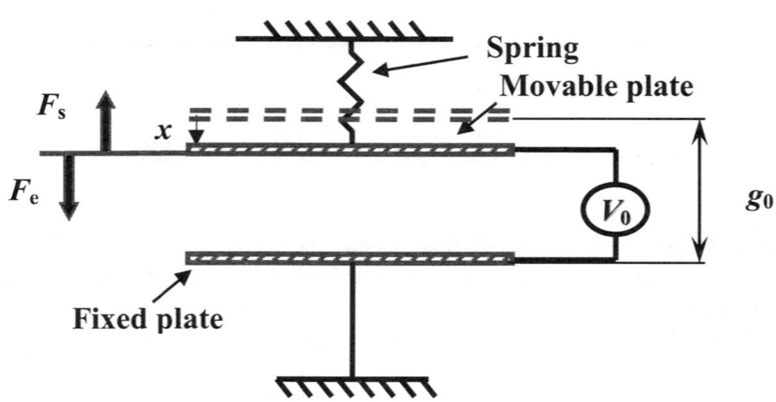


**Figure 2.** Single degree of freedom electromechanical system: static case.

At first it could be useful to consider the two elements (spring and capacitive actuator) of the electromechanical system separately (see Figure 3).

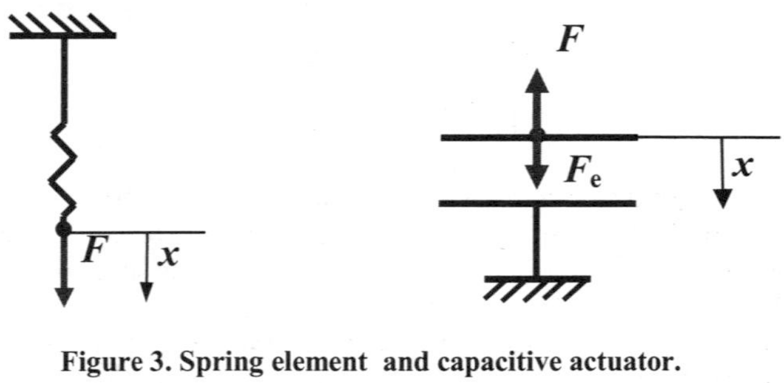


**Figure 3. Spring element and capacitive actuator.**

[1] Obviously the system is characterized by two degrees of freedom, if considered as electromechanical

The elastic strain energy $U_s$ stored by the spring is:

$$U_s = \frac{1}{2}kx^2 \tag{2-1}$$

The obvious equilibrium equation:

$$F = kx, \tag{2-2}$$

can be obtained also from the virtual work principle:

$$\delta U_s = \delta L \Rightarrow \delta\left(\frac{1}{2}kx^2\right) = F\delta x \Rightarrow kx = F \tag{2-3}$$

or, according to Castigliano theorem:

$$F = \frac{\partial U_s}{\partial x} = kx \tag{2-4}$$

The elastic strain energy $U_s$ stored by the spring is:

$$U_s = \frac{1}{2}kx^2 \tag{2-5}$$

For the capacitive actuator (see Figure 4) a similar approach can be followed; in the case of parallel plate capacitive actuator, the electrical characterization is missing. It is therefore necessary to specify if this is the case of an open circuit, for which the electrical charge $Q$=const., or if capacitor plates are connected to a voltage generator, i.e. $V=V_0$:

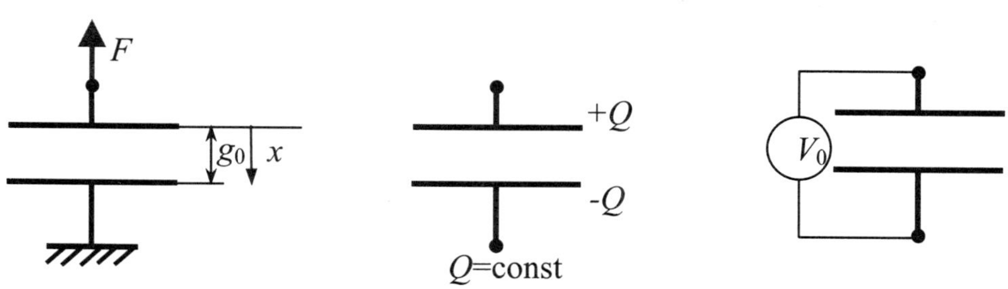


**Figure 4.** Capacitive actuator models: general scheme, open circuit or voltage generator.

The general expression of the energy stored by the capacitor is:

$$U_e = \frac{1}{2}VQ = \frac{1}{2}V^2C = \frac{1}{2}\frac{Q^2}{C} \tag{2-6}$$

where: $C = \frac{Q}{V} = \frac{\varepsilon A}{(g_0 - x)}$

$A$=parallel plate surface

$\varepsilon = \varepsilon_r\varepsilon_0$= permittivity of the dielectric.

## 2.1 Capacitive Actuator: Open Circuit

The energy stored by the actuator can be expressed as:

$$U_e = \frac{1}{2}\frac{Q^2}{C} = \frac{1}{2}Q^2\frac{(g_0 - x)}{\varepsilon A} \tag{2-7}$$

and therefore it is possible to evaluate the expression of the electrostatic force following a procedure similar to the case of the spring; in fact also in this case a virtual work principle holds and therefore:

$$\delta U_e = \delta L \Rightarrow -\frac{1}{2}\frac{Q^2}{\varepsilon A}\delta x = -F\delta x \Rightarrow F = \frac{1}{2}\frac{Q^2}{\varepsilon A} \tag{2-8}$$

or: $$F_e = -\frac{\partial U_e}{\partial x} = \frac{1}{2}\frac{Q^2}{\varepsilon A}. \tag{2-9}$$

It can be noticed that, for a positive small displacement increment $\delta x$ the stored potential energy decreases.

## 2.2 Capacitive Actuator: Constant Voltage Generator

The energy stored by the actuator can be expressed as:

$$U_e = \frac{1}{2}V_0Q = \frac{1}{2}V_0^2C = \frac{1}{2}\frac{\varepsilon A}{(g_0 - x)}V_0^2 \tag{2-10}$$

In this case: $$\frac{\partial U_e}{\partial x} = \frac{1}{2}\frac{\varepsilon A}{(g_0 - x)^2}V_0^2 > 0 \tag{2-11}$$

this means that the stored potential energy increases for a positive small displacement increment $\delta x$. This behaviour can be explained considering that in the case of a positive small displacement increment $\delta x$ , the voltage generator supplies a contribution $V_0 \delta Q$ that will be partially stored by the capacitor to satisfy the increasing storing energy requirement, and partially converted in mechanical work.

For a positive small displacement increment $\delta x$ the principle of energy conservation holds:

$$\delta U_e - \delta Q V_0 = \delta L \tag{2-12}$$

and it is therefore possible to write:

$$\delta(U_e - QV_0) = \delta L \Rightarrow \frac{1}{2}\frac{\varepsilon A}{(g_0 - x)} V_0^2 \delta x - \frac{\varepsilon A}{(g_0 - x)} V_0^2 \delta x = -F\delta x \tag{2-13}$$

$$-\frac{1}{2}\frac{\varepsilon A}{(g_0 - x)^2} V_0^2 \delta x = -F\delta x \Rightarrow F = \frac{1}{2}\frac{\varepsilon A}{(g_0 - x)^2} V_0^2 \tag{2-14}$$

or:

$$F = -\frac{\partial}{\partial x}(U_e - QV_0) = \frac{1}{2}\frac{\varepsilon A}{(g_0 - x)^2} V_0^2 . \tag{2-15}$$

That it is again a force that tends to attract the two plates. The above discussion can be resumed as shown in Figure 5.

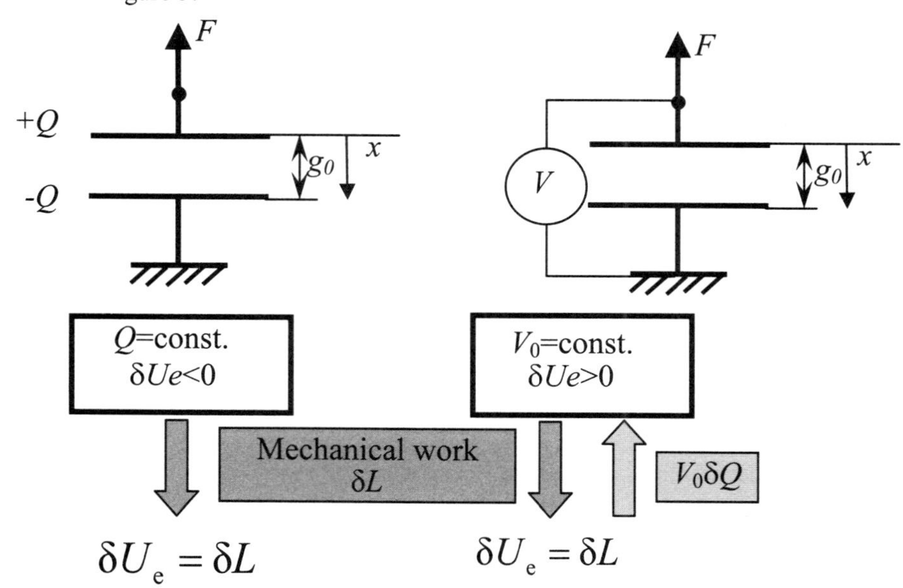


**Figure 5.** Energy balance for the electrostatic actuator (open circuit and voltage generator).

## 2.3 Single Degree of Freedom Electromechanical System: Open Circuit

Now the case of a *"single"* degree of freedom electromechanical system can be analyzed (see Figure 6); it possible to consider at first the case where the actuators plates are charged but not connected to a generator (open circuit).

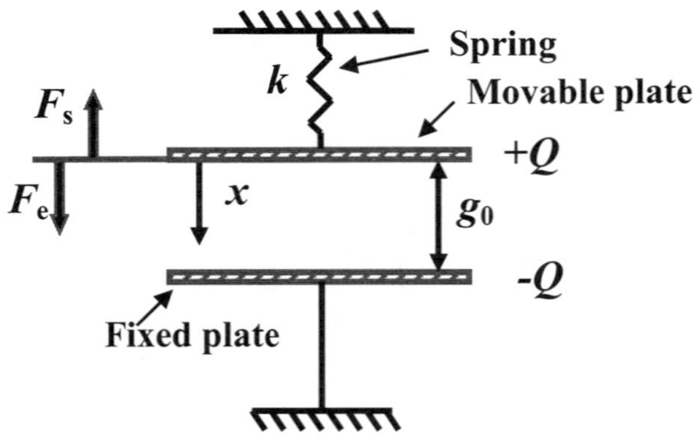


**Figure 6.** "Single" degree of freedom electromechanical system: open circuit.

If the displacement $x$ and the potential $V$ are chosen as mechanical and electrical generalized coordinates, the "total potential energy" can be expressed as:

$$U = \frac{1}{2}\begin{Bmatrix} x \\ V \end{Bmatrix}^T \begin{bmatrix} k & 0 \\ 0 & C \end{bmatrix} \begin{Bmatrix} x \\ V \end{Bmatrix} = \frac{1}{2}kx^2 + \frac{1}{2}CV^2 \tag{2-16}$$

It follows that the mechanical equilibrium in $x$ direction can be easily obtained from virtual work principle:

$$\delta U = \delta L = 0 \tag{2-17}$$

for this purpose it is necessary to introduce the electrical constitutive equation:

$$Q = CV \tag{2-18}$$

therefore it follows that:

$$U = \frac{1}{2}kx^2 + \frac{1}{2}\frac{Q^2}{C} \tag{2-19}$$

(2-20)

and:
$$kx\delta x - \frac{1}{2}\frac{Q^2}{\varepsilon A}\delta x = 0 \Rightarrow \quad kx - \frac{1}{2}\frac{Q^2}{\varepsilon A} \tag{2-21}$$

in conclusion:
$$F_s = F_e \tag{2-22}$$

where: $F_s = kx$ is the elastic restoring force of the spring

and: $F_e = \frac{1}{2}\frac{Q^2}{\varepsilon A}$ is the electrostatic force attracting the plates.

In matrix form, the governing equations of the problem, for a given electrical charge $Q$, are:

$$\begin{bmatrix} k & 0 \\ 0 & C \end{bmatrix}\begin{Bmatrix} x \\ V \end{Bmatrix} = \begin{Bmatrix} \frac{1}{2}\frac{Q^2}{\varepsilon A} \\ Q \end{Bmatrix} \tag{2-23}$$

It can be noticed that this choice of generalized coordinates is coherent with the formalism of considering them as the "unknown variables" of the two problems, and consequently the forcing terms (respectively the force and the charge) appear in the right side of the governing equations.

## 2.4 Single Degree of Freedom Electromechanical System: Voltage Generator

Now the case of a "single" degree of freedom electromechanical system controlled in voltage represented in Figure 7 will be considered.

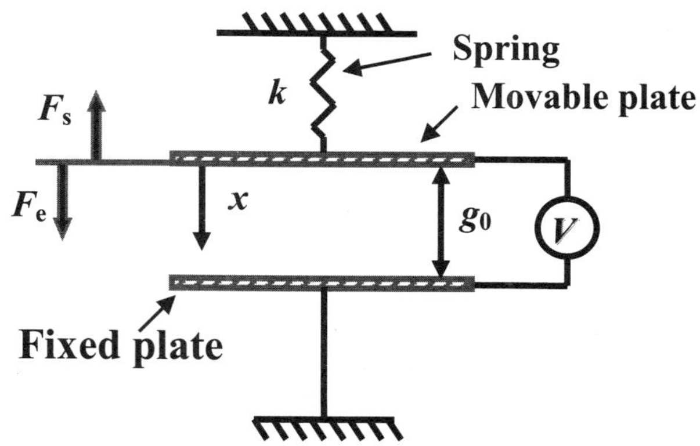


**Figure 7.** Single degree of freedom electromechanical system: voltage generator.

The expression of the total energy stored by the spring and the capacitor is:

$$U = \frac{1}{2}\begin{Bmatrix} x \\ V \end{Bmatrix}^T \begin{bmatrix} k & 0 \\ 0 & C \end{bmatrix} \begin{Bmatrix} x \\ V \end{Bmatrix} = \frac{1}{2}kx^2 + \frac{1}{2}V^2 C = \frac{1}{2}kx^2 + \frac{1}{2}V^2 \frac{\varepsilon A}{(g_0 - x)} \tag{2-24}$$

It can be noticed that also in this case the generalized coordinate of the electrical domain is voltage $V$. This choice is generally followed by Finite Element commercial codes.

Mechanical equilibrium in $x$ direction can be obtained following the usual procedure:

$$\delta U - V\delta Q = \delta L = 0 \tag{2-25}$$

as:

$$U = \frac{1}{2}kx^2 + \frac{1}{2}V^2 \frac{\varepsilon A}{(g_0 - x)} \tag{2-26}$$

it follows that:

$$kx\delta x + \frac{1}{2}\frac{\varepsilon A}{(g_0 - x)^2}V^2\delta x - \frac{\varepsilon A}{(g_0 - x)^2}V^2\delta x = 0 \tag{2-27}$$

in conclusion:

$$kx - \frac{1}{2}\frac{\varepsilon A}{(g_0 - x)^2}V^2 = F_s - F_e = 0 \tag{2-28}$$

where: $F_s = kx$ is the elastic restoring force of the spring

$F_e = \frac{1}{2}\frac{\varepsilon A}{(g_0 - x)^2}V^2$ is the capacitor electrostatic force of attraction

Following a matrix notation, the electromechanical equilibrium equations of the problem, in the case of a given voltage $V$, are:

$$\begin{bmatrix} k & 0 \\ 0 & C \end{bmatrix} \begin{Bmatrix} x \\ V \end{Bmatrix} = \begin{Bmatrix} \frac{1}{2}\frac{\varepsilon A}{(g_0 - x)^2}V^2 \\ Q \end{Bmatrix} \tag{2-29}$$

It can be enhanced again that, in this case, the choice of the generalized electrical coordinate $V$ seems less natural, as physically it is the "electrical" forcing term.

For a better physical understanding of the electro- mechanical behaviour of the actuator it could be useful to choose the charge $Q$ as electrical generalized coordinate. If the electrostatic force is expressed as: $F_e = \frac{1}{2}\frac{\varepsilon A}{(g_0 - x)^2}V^2 = \frac{1}{2}\frac{Q^2}{\varepsilon A}$ and the capacitance as: $C = \frac{\varepsilon A}{(g_0 - x)}$

the electromechanical equations of the system are:

$$\begin{bmatrix} k & 0 \\ 0 & \frac{(g_0 - x)}{\varepsilon A} \end{bmatrix} \begin{Bmatrix} x \\ Q \end{Bmatrix} = \begin{Bmatrix} \frac{1}{2} \frac{Q^2}{\varepsilon A} \\ V \end{Bmatrix} \tag{2-30}$$

This equations clearly shows that there is a coupling between the electrical and the mechanical domain; the second equation permits to obtain for a chosen position of the actuator $x$ and for an applied voltage $V$ the value of the electrical charge $Q$. Introducing $Q$ in the second term of the first equation permits the electrostatic force to be computed and consequently a new displacement $x$ can be evaluated.

The described approach can be used to find the solution iteratively following a so called sequential field coupling method. As it will be described successively, this approach is commonly used to solve the case of multi-degree of freedom systems; on the other hand in this case the mechanical solution can be directly obtained by substitution, in fact:

$$F_s(x) - F_e(x) = 0 \Rightarrow kx - \frac{1}{2} \frac{\varepsilon A}{(g_0 - x)^2} V^2 = 0 \tag{2-31}$$

The solution can be easily obtained graphically if $F_s(x)$ and $F_e(x)$ are plotted separately and their crossing point considered.

In dimensionless form:

$$\zeta - \frac{4}{27} \frac{v^2}{(1-\zeta)^2} = 0 \tag{2-32}$$

where: $\zeta = \dfrac{x}{g_0}$ $\quad v = V / V_{\text{pull-in}}$ $\quad V_{\text{pull-in}} = \sqrt{\dfrac{8}{27} \dfrac{g_0 k}{\varepsilon A}}$

Physically, this is again an equilibrium equation, in fact, putting:

$f_s = \zeta$ (expression corresponding to dimensionless spring force), and:

$f_e = \dfrac{4}{27} \dfrac{v^2}{(1-\zeta)^2}$ (expression corresponding to a dimensionless electrostatic force),

It follows:

$$\zeta = \frac{4}{27} \frac{v^2}{(1-\zeta)^2} \Rightarrow f_s = f_e \tag{2-33}$$

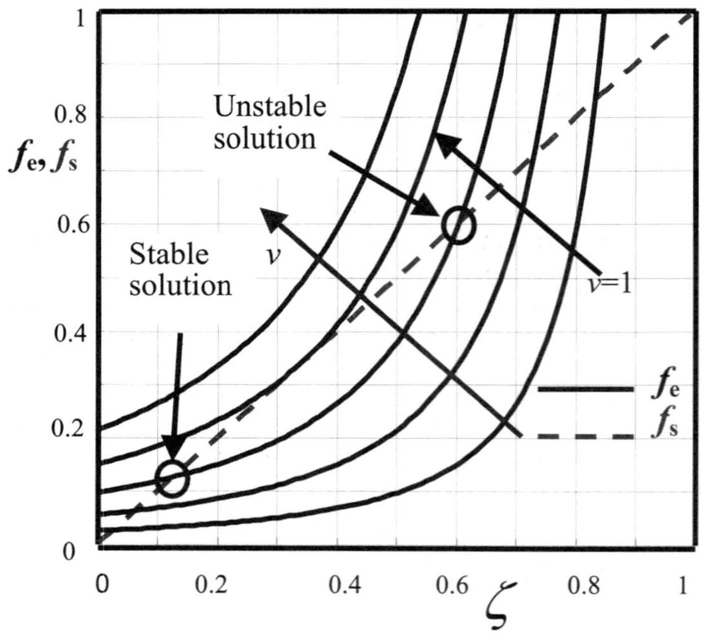


**Figure 8.** Stable and unstable equilibrium configurations.

It seems that:

$$kx - \frac{1}{2}\frac{\varepsilon A}{(g_0 - x)^2}V^2 = 0 \tag{2-34}$$

was an algebraic third degree equation and therefore it should have one, or three real solutions. In reality the total potential energy has the expression:

$$U = \frac{1}{2}\begin{Bmatrix} x \\ V \end{Bmatrix}^T \begin{bmatrix} k & 0 \\ 0 & C \end{bmatrix}\begin{Bmatrix} x \\ V \end{Bmatrix} = \frac{1}{2}kx^2 + \frac{1}{2}V^2C = \frac{1}{2}kx^2 + \frac{1}{2}V^2\frac{\varepsilon A}{|g_0 - x|} \tag{2-35}$$

and therefore the above equation becomes:

$$kx - \frac{1}{2}\frac{\varepsilon A}{(g_0 - x)^2}V^2 = 0, \quad x < g_0 \quad \text{with two solutions and:} \tag{2-36}$$

$$kx + \frac{1}{2}\frac{\varepsilon A}{(g_0 - x)^2}V^2 = 0, \quad x > g_0 \quad \text{with no solution.} \tag{2-37}$$

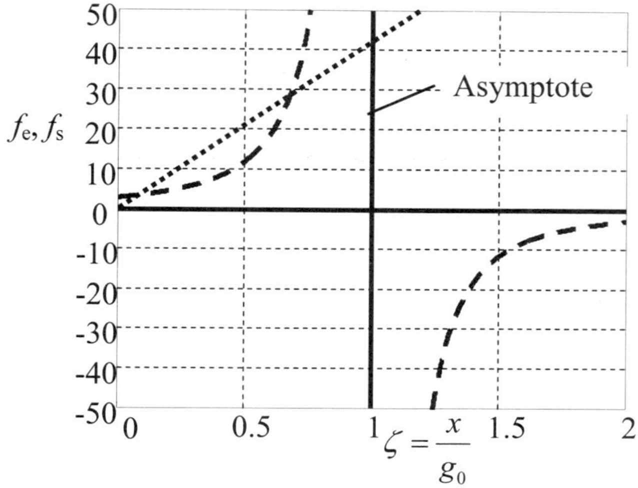


| $kx - \frac{1}{2}\frac{\varepsilon A}{(g_0 - x)^2}V^2 = 0, \quad x < g_0$ | $kx + \frac{1}{2}\frac{\varepsilon A}{(g_0 - x)^2}V^2 = 0, \quad x > g_0$ |
|---|---|

**Figure 9.** Physical and "not-physical" equilibrium configurations.

It can be noticed that for voltages higher than a value called $V_{\text{pull-in}}$, no equilibrium is possible. For values lower than the $V_{\text{pull-in}}$ two solutions, one stable and one unstable exist.

In order to study the stability of the mechanical equilibrium, it is necessary to refer to:

$$\delta(\delta U - V\delta Q) = \delta\left(kx\delta x - \frac{1}{2}\frac{\varepsilon A}{(g_0 - x)^2}V^2\delta x\right) = \left(k - \frac{\varepsilon A}{(g_0 - x)^3}V^2\right)\delta x^2 \tag{2-38}$$

Putting: $\Phi = k - \frac{\varepsilon A}{(g_0 - x)^3}V^2$,

stability can be studied considering the sign of $\Phi$ and remembering:

$$kx - \frac{1}{2}\frac{\varepsilon A}{(g_0 - x)^2}V^2 = 0 \tag{2-39}$$

An other possible approach to explore pull-in is to make a graph of the normalized displacement $(1-\zeta)=\left(1-\dfrac{x}{g_0}\right)$ versus the normalized voltage $v=\dfrac{V}{V_{\text{pull-in}}}$.

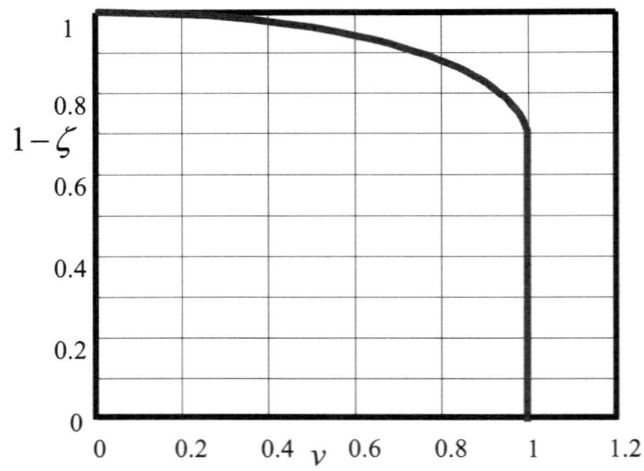


**Figure 10.** Normalized displacement versus normalized voltage.

It can be noticed that, when $x$ reaches one third of the gap, unstable collapse to 0 gap occurs, the corresponding abscissa is pull-in normalized voltage. The previous representation can be misleading, if the mechanical non-linearity has to be enhanced. In fact, as the electrostatic force is proportional to $V^2$, the relationship between the movable plate displacement and the square of the applied voltage represented in Fig.11 shows clearly that non-linearity is significant only for $v^2 > 0.6$ corresponding to a displacement above 10 %.

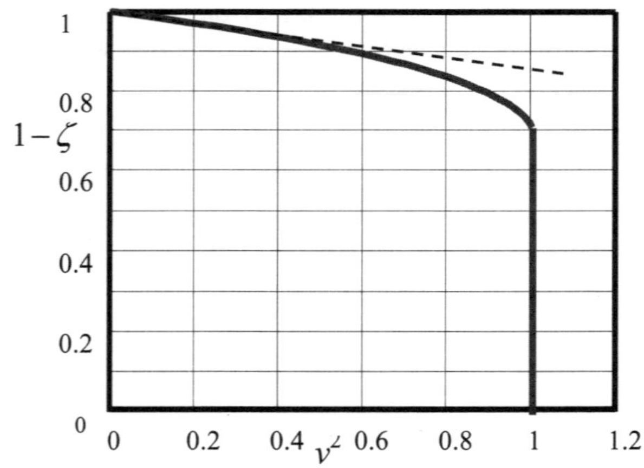


**Figure 11.** Normalized displacement versus squared normalized voltage.

# 3 Multi-Degree of Freedom System

In the case of structures which undergo elastic deformations the problem in general requires a numerical approach. By resorting to Finite Element models (FEM) for the mechanical domain and FEM or boundary element models (BEM) for the electrical domain, it is generally possible to discretize the problem obtaining a non linear system of coupled algebraic equations.

In the general case of a problem characterized by two physical domains, and therefore characterized by two types of generalized coordinates $\{x\}$, $\{y\}$ the governing equations of the discretized system are:

$$\begin{bmatrix} [k_{11}] & [k_{12}] \\ [k_{21}] & [k_{22}] \end{bmatrix} \begin{Bmatrix} \{x\} \\ \{y\} \end{Bmatrix} = \begin{Bmatrix} \{F_x\} \\ \{F_y\} \end{Bmatrix} \tag{3-1}$$

If all the coefficients of the system of equations are constant, the system can be called linearly coupled. A well know example of linearly coupled electromechanical system is the case of piezoelectric element (Bona et al., 1994). If the static equilibrium of a piezoelectric element is considered, the following equations can be written:

$$\begin{bmatrix} [K] & [\Theta] \\ [\Theta^T] & [C_p] \end{bmatrix} \begin{Bmatrix} \{x\} \\ \{V\} \end{Bmatrix} = \begin{Bmatrix} \{F\} \\ \{Q\} \end{Bmatrix} \tag{3-2}$$

being:

$K$ : stiffness matrix
$\Theta$ : piezoelectric coefficient coupling matrix
$C_P$: piezoelectric capacitance matrix

It must be noticed that in this case coupling effect is electromechanical, i.e. coordinates are selected among mechanical and electrical degrees of freedom, but, with respect to electrostatic actuation, it is applied to the overall volume of the piezoelectric material. In the piezoelectric case the coupling is due only to the coupling matrix $\Theta$, and it affects exclusively the elements of the matrix out of the main diagonal. Because all stiffness matrix elements and forces do not depend on the system coordinates, the coupling can be considered linear and solution is quite straightforward since for given forces $F$ and charge $Q$, displacement $x$ and voltage V can be easily computed.

In a more general case the system of equations can be "non linearly" coupled, as the coefficient of the system depends on the system variables, a possible configuration could be the following:

$$\begin{bmatrix} [k_{11}(x,y)] & [k_{12}(x,y)] \\ [k_{21}(x,y)] & [k_{22}(x,y)] \end{bmatrix} \begin{Bmatrix} \{x\} \\ \{y\} \end{Bmatrix} = \begin{Bmatrix} \{F_x(x,y)\} \\ \{F_y(x,y)\} \end{Bmatrix} \tag{3-3}$$

considering that $[k_{12}(x,y)]$ and $[k_{21}(x,y)]$ elements can be always moved to the right side of the system:

$$\begin{bmatrix} [k_{11}(x,y)] & [0] \\ [0] & [k_{22}(x,y)] \end{bmatrix} \begin{Bmatrix} \{x\} \\ \{y\} \end{Bmatrix} = \begin{Bmatrix} \{\overline{F}_x(x,y)\} \\ \{\overline{F}_y(x,y)\} \end{Bmatrix} \tag{3-4}$$

It can be noticed that coupling still applies, although matrix elements $[k_{12}(x,y)]$ and $[k_{21}(x,y)]$ do not appear, since $[k_{11}(x,y)]$ and $[k_{22}(x,y)]$ elements and forces at right hand of the equation depend on the system coordinates.

In particular, in the case of a system discretized following a variational approach (es. FEM) the following system of equations can be obtained:

$$\begin{bmatrix} [k_{11}(x,y)] & [0] \\ [0] & [k_{22}(x,y)] \end{bmatrix} \begin{Bmatrix} \{\Delta x\} \\ \{\Delta y\} \end{Bmatrix} = \begin{Bmatrix} \{\Delta F_x(x,y)\} \\ \{\Delta F_y(x,y)\} \end{Bmatrix} \tag{3-5}$$

where increments $\Delta$ are introduced.

In the case of an electrostatically actuated system a simpler set of equations is generally derived:

$$\begin{bmatrix} [k_{11}] & [0] \\ [0] & [k_{22}(x)] \end{bmatrix} \begin{Bmatrix} \{x\} \\ \{y\} \end{Bmatrix} = \begin{Bmatrix} \{F_x(y)\} \\ \{F_y)\} \end{Bmatrix} \tag{3-6}$$

If the usual FEM approach is applied to the electrical and to the mechanical domains, the following system of equations is obtained:

$$\begin{bmatrix} [k] & [0] \\ [0] & [C(u)] \end{bmatrix} \begin{Bmatrix} \{u\} \\ \{V\} \end{Bmatrix} = \begin{Bmatrix} \{F(V)\} \\ \{Q\} \end{Bmatrix} + \begin{Bmatrix} \{F\} \\ \{0\} \end{Bmatrix} \tag{3-7}$$

where $F(V)$ is the electromechanical force.

**Table 2.** Comparison between thermal and electrostatic domains: governing equation.

| **Thermal problem** | **Electrostatic problem** |
|---|---|
| $k_T \Delta T = -F$<br>in 2D: $k\left(\frac{\partial^2 T}{\partial x^2} + \frac{\partial^2 T}{\partial y^2}\right) = -F$<br>$F$: heat load vector, [W/m$^3$] | $\varepsilon \Delta V = -\rho$<br>in 2D: $\varepsilon\left(\frac{\partial^2 V}{\partial x^2} + \frac{\partial^2 V}{\partial y^2}\right) = -\rho$<br>$\rho$: charge density, [C/m$^3$] |

**Table 3.** Comparison between thermal and electrostatic domains: FEM discretisation.

<table>
<tr><th colspan="2">Finite element method</th></tr>
<tr><th>Thermal problem</th><th>Electrostatic problem</th></tr>
<tr><td>Nodal element temperature:<br>$\{T\}=\begin{Bmatrix}T_i\\T_j\\T_k\end{Bmatrix}\Rightarrow T(x,y)=[N(x,y)]\{T\}$</td><td>Nodal element potential:<br>$\{V\}=\begin{Bmatrix}V_i\\V_j\\V_k\end{Bmatrix}\Rightarrow V(x,y)=[N(x,y)]\{V\}$</td></tr>
<tr><td colspan="2">➔ the same interpolation function $[N(x,y)]$</td></tr>
<tr><td>$\{\varepsilon^*\}=\begin{Bmatrix}\frac{\partial T}{\partial x}\\\frac{\partial T}{\partial y}\end{Bmatrix}\Rightarrow\{\varepsilon^*(x,y)\}=[B(x,y)]\{T\}$<br>-grad(T) ➔heat flux</td><td>$\{\varepsilon^*\}=\begin{Bmatrix}\frac{\partial V}{\partial x}\\\frac{\partial V}{\partial y}\end{Bmatrix}\Rightarrow\{\varepsilon^*(x,y)\}=[B(x,y)]\{V\}$<br>-grad(V) (electric field)</td></tr>
<tr><td colspan="2">➔ the same matrix $[B(x,y)]$ units:[m$^{-1}$]</td></tr>
<tr><td>$[D]=\begin{bmatrix}k&0\\0&k\end{bmatrix}$<br>$K_T$ – thermal conductivity coefficient, [W/Km]</td><td>$[D]=\begin{bmatrix}\varepsilon&0\\0&\varepsilon\end{bmatrix}$<br>$\varepsilon$ - permittivity coefficient, [F/m]</td></tr>
<tr><td colspan="2">$[k]=t\int_A[B]^T[D][B]dA$, $A$ is the finite element surface</td></tr>
<tr><td colspan="2">Assembling process</td></tr>
<tr><td colspan="2">$[K]=\sum_{n_{el}}[k]$</td></tr>
<tr><td>$[K]\{T\}=\{F\}$</td><td>$[K]\{V\}=\{Q\}$</td></tr>
<tr><td>$\{F\}=\sum_{n_{el}}\int_{V_{el}}[N]^T F dv$, [W]</td><td>$\{Q\}=\sum_{n_{el}}\int_{V_{el}}[N]^T \rho dv$, [C]</td></tr>
</table>

It can be observed that, similarly to the single degree of freedom system the coupling between the two domains is mainly due to the fact that the mechanical forces depend on the electrical variables, whereas the capacitance matrix terms depends on the generalized displacements $u$ that characterize the mechanical domain.

A well known dual case could be that of thermo-elasticity:

$$\begin{bmatrix}[k]&[0]\\[0]&[k_T]\end{bmatrix}\begin{Bmatrix}\{u\}\\\{T\}\end{Bmatrix}=\begin{Bmatrix}\{F(T)\}\\\{Q_T\}\end{Bmatrix}+\begin{Bmatrix}\{F\}\\0\end{Bmatrix} \quad (3\text{-}8)$$

where: $k_T$=conductivity matrix
$Q_T$= heat vector
$F(T)$=heat load vector

As shown in Table 2, this case is really dual in the sense that the governing equation (Poisson's equation) is the same for the electrical and for the thermal domains. On the other hand in this case the temperature distribution produces generally small displacements $u$ and therefore the conductivity matrix $k_T$ terms remain practically unchanged. Therefore the well known procedure of calculating the temperature distribution from the second set of equations and the mechanical displacements from the first ones can be followed.

If the electrostatic case is considered again, as shown in Table 3, the FEM approach that must be followed obviously is perfectly dual with respect to the thermal case.

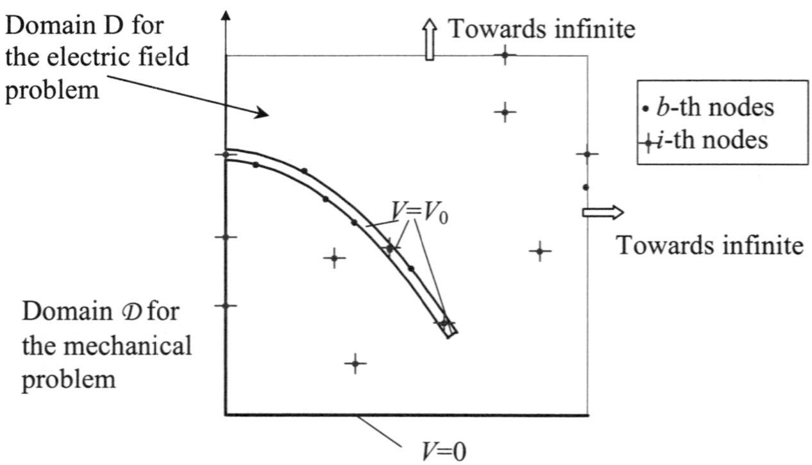


**Figure12.** Nodes at the interface between mechanical and electrical domains.

When the two domains are discretized, a simplification follows from the consideration that the two domains (mechanical and electrical) interfere each others only at the boundary (such kind of coupling is also called "surface coupling"), Figure 13. Therefore the governing equations can be written partitioning the degrees of freedom of the nodes at the boundary between the two domains (subscript $b$ ) and those which are outside this boundary (subscript $i$):

$$\begin{bmatrix} [k_{ii}] & [k_{ib}] & [0] & [0] \\ [k_{bi}] & [k_{bb}] & [0] & [0] \\ [0] & [0] & [C_{ii}(u_b)] & [C_{ib}(u_b)] \\ [0] & [0] & [C_{bi}(u_b)] & [C_{bb}(u_b)] \end{bmatrix} \begin{Bmatrix} \{u_i\} \\ \{u_b\} \\ \{V_i\} \\ V_0\{I\} \end{Bmatrix} = \begin{Bmatrix} \{0\} \\ \{F_{eb}(\{V_i\}, V_0)\} \\ \{0\} \\ \{Q\} \end{Bmatrix} \tag{3-9}$$

In consideration of the particular kind of coupling that characterizes the problem, solution is no more so straightforward: it involves the external environment surrounding the investigated structure, while in the previous example of the piezoelectric the volume interested coincided with the piezoelectric material. So far a dedicated technique has to be applied, a usual approach is the so called "sequential coupling" approach:

-for a certain value of the displacements $u$ ($u$=0, at the first iteration, corresponding to the undeformed structure) the electrical problem:

$$\begin{bmatrix} [C_{\mathrm{ii}}(u_{\mathrm{b}})] & [C_{\mathrm{ib}}(u_{\mathrm{b}})] \\ [C_{\mathrm{bi}}(u_{\mathrm{b}})] & [C_{\mathrm{bb}}(u_{\mathrm{b}})] \end{bmatrix} \begin{Bmatrix} \{V_{\mathrm{i}}\} \\ V_0\{\mathrm{I}\} \end{Bmatrix} = \begin{Bmatrix} \{0\} \\ \{Q\} \end{Bmatrix} \tag{3-10}$$

is solved and the unknown voltage $\{V_{\mathrm{i}}\}$ is determined for a value of the applied voltage $V_0$ from the first set of equations:

$$\{V_{\mathrm{i}}\} = -V_0[C_{\mathrm{ii}}(u_{\mathrm{b}})]^{-1}[C_{\mathrm{ib}}(u_{\mathrm{b}})] \tag{3-11}$$

The corresponding value of the electrostatic field $E$ and of the electrostatic loads $F_{\mathrm{eb}}$ applied at the interface can thus be determined.

For the $j$-th electrostatic finite element (the hypothesis of a FEM discretization is done) positioned at the boundary, it holds:

$$V(x,y) = [N(x,y)]_j\{V\}_j \tag{3-12}$$

where $[N(x,y)]$ is the element interpolation function.

The electric field inside the $j$-th element is given by the relation:

$$\{E\} = \begin{Bmatrix} E_x(x,y) \\ E_y(x,y) \end{Bmatrix} = \begin{bmatrix} \dfrac{\partial N}{\partial x} \\ \dfrac{\partial N}{\partial y} \end{bmatrix}_j \{V\}_j = [B(x,y)]_j\{V\}_j \tag{3-13}$$

As:

$$\{V\}_j = \begin{Bmatrix} \{V_i\}_j \\ \{V_b\}_j \end{Bmatrix} = V_0 \begin{Bmatrix} -\{[C_{\mathrm{ii}}(u_{\mathrm{b}})]^{-1}[C_{\mathrm{ib}}(u_{\mathrm{b}})]\}_j \\ \{\mathrm{I}\} \end{Bmatrix} \tag{3-14}$$

$$\{E\} = V_0[B(x,y)]_j \begin{Bmatrix} -\{[C_{\mathrm{ii}}(u_{\mathrm{b}})]^{-1}[C_{\mathrm{ib}}(u_{\mathrm{b}})]\}_j \\ \{\mathrm{I}\} \end{Bmatrix} = V_0\{f(u_b)\}_j \tag{3-15}$$

The distribution of the electromechanical forces, produced by the electric field and acting on the boundary, can be computed by means of the Maxwell stress tensor $[\sigma]$, for the $j$-th element:

$$
\begin{aligned}
[\sigma] &= \varepsilon\left(\{E\}\{E\}^T - \frac{1}{2}\{E\}^T\{E\}[\mathrm{I}]\right) = \\
&= \varepsilon V_0^2\left(\{f(u_b)\}_j\{f(u_b)\}_j^{\,T} - \frac{1}{2}\{f(u_b)\}_j^{\,T}\{f(u_b)\}_j[\mathrm{I}]\right) = \varepsilon V_0^2[G(u_b)]
\end{aligned}
\tag{3-16}
$$

Finally the forces acting at the interface nodes $b$ can be computed:

$$
\{F_{\mathrm{eb}}\} = \int_{S_{\mathrm{b}}}[\sigma]\{n\}dS_{\mathrm{b}} = \varepsilon V_0^2\int_{S_{\mathrm{b}}}[G(u_{\mathrm{b}})]\{n\}dS_{\mathrm{b}}
\tag{3-17}
$$

where $\{n\}$ is the surface normal unit vector.

The new displacements $\{u_{\mathrm{i}}\}$ and $\{u_{\mathrm{b}}\}$ are then evaluated solving the mechanical problem:

$$
\begin{bmatrix} [k_{\mathrm{ii}}] & [k_{\mathrm{ib}}] \\ [k_{\mathrm{bi}}] & [k_{\mathrm{bb}}] \end{bmatrix}\begin{Bmatrix} \{u_{\mathrm{i}}\} \\ \{u_{\mathrm{b}}\} \end{Bmatrix} = \begin{Bmatrix} \{0\} \\ \{F_{eb}\} \end{Bmatrix}
\tag{3-18}
$$

For the new values of $\{u_{\mathrm{b}}\}$an updated electrical domain is defined, a new mesh is generated and the electrical problem is solved again. The procedure is repeated until the difference between the mechanical displacements of two subsequent iterations is under a certain tolerance.

Finally, it has to be noticed that at each iteration the electrical domain changes its boundary and therefore needs to be re-meshed, on the other hand the mechanical domain mesh does not require to be modified. It is not absolutely necessary to have correspondence between the mechanical and the electrical mesh node at the interface; obviously, if the nodes have different positions, suitable interpolation function are necessary, to apply the electrical forces to the mechanical domain and to impose the corresponding displacements to the electrical domain interface.

## References

Bona, B., Brusa, E., Canestrelli, P., Genta, G. and Tonoli, A. (1994), Finite element modeling and experimental validation of a elastic beam with surface bonded piezoelectric devices, Proc. IEEE Int. Conf. on Rob. and Automation, San Diego, May 1994, 2659-2664.

Fukuda, T. and Menz ,W., eds. (1998). *Handbook of Sensor and Actuators*, Vol.6, (Amsterdam), Elsevier.

Hsu Tai-Ran (2002). *MEMS & Microsystems; Design and Manufacture*, New York, Mc Graw Hill.

Madou, M. J., (2002), *Fundamentals of Microfabrication*, II Ed., Boca Raton, FL, CRC Press.

Dornheim, M., A., (1998), Tiny drones may be soldier's new tool, *Aviation Week&Space Technology*, 8:42-48.

Nguyen Clark, T-C, (2000). Micromechanical circuits for communication transceivers, *IEEE BCTM 8.2*, 142-149.

Senturia, S. D., (2001). *Microsystems Design*, Boston, MA, Kluwer Academic.

Trimmer W.S.N., (1989), Microrobots and micromechanical systems, Sensors and Actuators, 19, 167-287.

# Dynamics of Mechatronic Systems at Microscale

Eugenio G. M. Brusa

Dipartimento di Ingegneria Elettrica, Gestionale e Meccanica (DIEGM)
University of Udine, Italy

**Abstract.** Predicting dynamic behavior of nonlinear micromechatronic systems is rather difficult, because of the effects produced by electromechanical coupling, nonlinearities and microscale. This lecture is aimed to introduce basic methods for mechanical design of microelectromechanical systems (MEMs), by means of a test case consisting of an electrostatic microactuator, with few degrees of freedom. A model of the whole microsystem is built, by following a systematic approach, typical of structural mechatronics, dealing with formulation and solution of the equations of motion. Methods herewith presented are already extended to continuous microsystems, applied to numerical methods and implemented in some commercial and academic tools.

## 1 Introduction

Dynamic behaviour prediction is a relevant issue of mechanical design of microelectromechanical systems (MEMs), like microaccelerometers, radio-frequency devices (RF-MEMs), microswitches and microresonators (Senturia,2000; Giurgiutiu,2004). Working condition in the above microsystems consists of motion of either a compliant microstructure or a rigid body, suspended by elastic supports, in presence of electromechanical actuation. Geometries often proposed are microbeams, membranes (compliant) and microplates (rigid), while typical actuations are piezoelectric, electrostatic and magnetic. In practice, to select suitable material and fabrication process, designer is required to answer to few basic questions, concerning dynamics, for given frequency and actuation amplitude: what are maximum displacement and maximum rotation induced by vibration? Where are those located on the structure? Or even what is vibration mode of microsystem? Is it vibration's trend stable or unstable?

These questions actually require to resort to rather complicate models, to deal with all the effects present at microscale. To define outlines of this introduction, a brief overview of the main peculiarities of microscale are herewith summarized.

Microstructural system can be either continuous (beams, membranes) or discretely concentrated, like in case of single capacitors, with movable rigid plates, inductors or piezostacks. Continuous microsystems are modeled by resorting to numerical discretization techniques, like Finite Element Method (FEM), or by reducing the number of degrees of freedom, leading to equivalent lumped parameters models (Genta,2000; Meirovitch,1990).

Moreover actuation may be linear or nonlinear function of mechanical and electrical coordinates. Piezoelectric action is, for instance, linearly dependent on applied voltage, while electrostatic force is nonlinear in voltage, charge and displacement. Nonlinear actions require a numerical integration of equations of motion, by algorithms suitable to assure numerical stability and convergence (Cook,1989; Bathe,1996).

Electromechanical coupling can be interpreted either as result of a global effect of related electric or magnetic field, by computing the resulting electromechanical forces applied to the microsystem, or as locally distributed effect, corresponding to a load distribution on microstructure surface. While in the first case lumped parameters models describe actuation forces and moments, in the second case a detailed analysis is performed by discretization techniques, like FEM, as for structural domain, or Boundary Element Method (BEM).

In addition to the above mentioned issues, a mechatronic approach takes into account that microsystem is part of an electric network, and dynamic behavior has to be evaluated in connection with electronic circuit (Meirovitch,1990; Rizzoni,2000). Moreover, microstructure is very often an actively controlled system more than an uncontrolled device vibrating within the electric field.

All the above issues motivate current research activity about MEMs dynamics modeling and characterization (Shi,1996; Ananantasuresh,1996; Younis,2003). These outlines focus design aspects applicable to the simple case of nonlinear electrostatic microactuator, consisting of a single degree of freedom capacitor, connected to a vibrating mass, suspended by an elastic spring. Although so simple, this microsystem is suitable to introduce approaches applied by numerical methods in analyzing continuous and controlled systems. Analysis focuses the electromechanical coupling between mechanical microstructure and electric actuator, by providing dynamic equilibrium equations for an integrated design of both the microstructure and the connected electric circuit.

# 2 Formulation of the dynamic equilibrium equations of the coupled microsystem

## 2.1 Applicable approaches

A first task of design procedure consists of the formulation of the equations of motion of the analysed microsystem. Two very popular approaches are D'Alembert's and Lagrange's formulations. Since '60s several authors demonstrated that both can be generalized to the case of electromechanically coupled systems, with electric and mechanical d.o.f. (Crandall, Woodson, 1968; Meisel,1973; Meirovitch,1990). D'Alembert suggests of balancing mechanical, electrical and electromechanical actions, applied to the microsystem, while Lagrange's equations are obtained by computing energy and co-energy contributions, i.e., kinetic, elastic, electrical and magnetic to derive the so-called Lagrangian function. Both procedures are herewith applied to the electrostatic microactuator.

## 2.2 Description of test case: the electrostatic microactuator

Following sections deal with an electrostatic microactuator, depicted in figure 1, including: a capacitor, with capacitance $C$, two ideally parallel conductive plates, dielectric constant $\varepsilon$ and plates area $A$; a concentrated mass $m$ suspended by an elastic spring, whose stiffness be $k$. To describe both the mechanical and electric behaviors, a selection of coordinates is required. Mechanical engineer is prone to classify this device as "single d.o.f. mechanism", nevertheless capacitor plays an energy conversion between the electric and the mechanical domains, only if the connected electric circuit either provides or absorbs the corresponding energy contribution.

This effect introduces a so-called electromechanical coupling, between mechanical and electrical coordinates. Therefore two coordinates are needed. A first mechanical coordinate is displacement of mass $m$, along $x$-axis. Be point "O", origin of $x$-axis, position of mass $m$ for which: spring does not apply any elastic restoring force, capacitor plates are located at initial gap $g_0$. A second electric coordinate is chosen between applied voltage $v$ and imposed charge $q$, depending on the formulation preferred, for the electrical analysis of the device, or even on the control strategy implemented. A final set of two coordinates holds, including either $x$ and $v$ or $x$ and $q$.

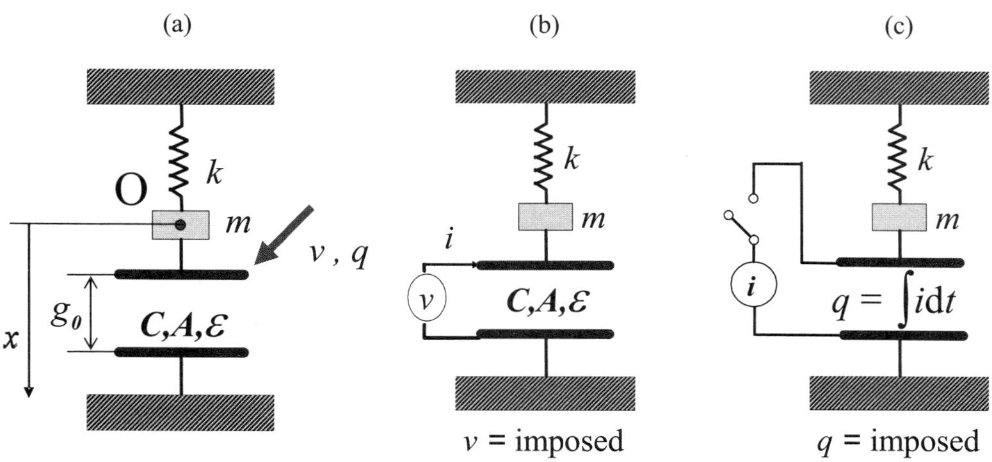


**Figure 1.** Ideal electrostatic microactuator (a), with imposed voltage (b) or imposed charge (c)

Next step in procedure defines the electromechanical constraints applied to the microsystem. Two kind of constraints are present: mechanical and electrical. Mechanical constraints are evidenced in figure 1: spring, mass and capacitive plates are connected as shown, in particular spring and capacitor are connected at one end to a fixed frame. Electrical constraints act in a such way that conductive plates can be either connected to an external circuit, in a closed loop, to a voltage generator or simply stand-alone, in an open loop, with no external power supplier. In practice, two configurations can be analyzed for the same device: (a) circuit with imposed voltage and variable charge; (b) circuit with imposed charge and variable voltage between plates. In figure 1b a total amount of charge $q$ is transferred on capacitor plates by the current generator, during a preliminary transitory, in which circuit is closed, then, when the required level of charge is achieved, circuit is opened and microcapacitor activated.

### 2.3 D'Alembert's generalized approach

If a pure balancing of applied electromechanical actions is followed to derive dynamic equilibrium equations of the whole microsystem, according to D'Alembert, and imposed voltage configuration is selected, contributions can be evaluated as follows (see figure 2): a spring elastic restoring force $F_s$, an inertial force $F_{in}$, a capacitor actuation $F_e$, attracting plates each other, either expressed in voltage or charge, and an additional external, mechanical, force $F$, if present:

$$F_s = -kx \qquad F_{in} = -m\ddot{x} \qquad F_e = \frac{1}{2}\frac{q^2}{\varepsilon A} \qquad F_e = \frac{1}{2}\frac{\varepsilon A v^2}{(g_0 - x)^2} \tag{2-1}$$

Above expressions of $F_e$ consider imposed charge and voltage respectively and apply both to constant and time variable actions: if charge is constant $q = Q = const.$, while if voltage is constant $v = V_0 = const.$

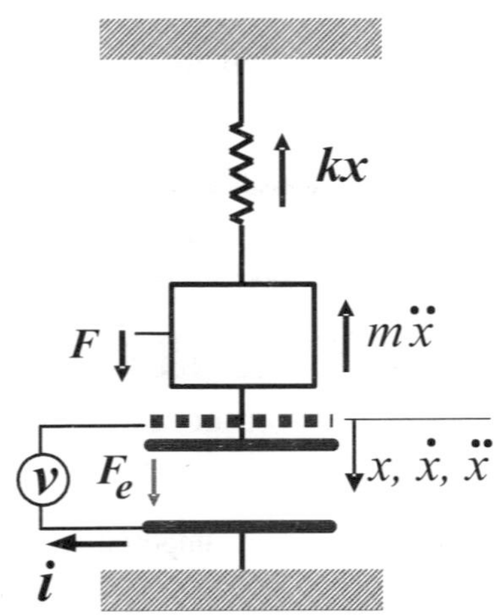


**Figure 2.** D'Alembert's dynamic equilibrium applied to the test case.

Above procedure leads to two formulations of the equation of motion, a first one expressed in charge coordinate:

$$m\ddot{x} + kx = F + \frac{\varepsilon A v^2}{2(g_0 - x)^2} \tag{2-2}$$

being voltage a forcing term at the right-hand of the above equation, while the same configuration is described by a second equation, in which voltage is coordinate and charge forcing term:

$$m\ddot{x} + kx = F + \frac{q^2}{2\varepsilon A} \tag{2-3}$$

It can be evidenced that in present case electromechanical actuator is approximated by an ideal capacitor, having no electrical resistance nor inductance. No dissipation is considered at this level, dynamic equilibrium is therefore established among known force $F$, electromechanical action $F_e$, inertial force $F_{in}$ and spring restoring force $F_s$.

A key remark on above equation is asking whether it is exhaustive in terms of description of dynamic behavior of the microsystem or not. Actually, it does not. Since microsystem is no more mechanical, but micromechatronic or microelectromechanical, a complete and exhaustive prediction can be drawn only if a similar equation is written for the electric behavior. In fact, electric circuit exhibits a dynamic behavior too, depending on the variability of capacitance in time. In network analysis, an approach similar to the D'Alembert's, in mechanics, can be followed to derive a second equation, according to Kirchoff's rules of loop and nodes equilibrium. The latter can be written either for open or closed circuit, corresponding to the two above defined constraints of imposed $q$ or $v$. Forcing terms are voltage and current respectively.

Two equations hold, a first is:

$$q\frac{g_0 - x}{\varepsilon A} = v \tag{2-4}$$

in which charge $q$ is coordinate and voltage $v$ is the corresponding electric force; actually it corresponds to the definition of capacitance in electrical engineering, and by converse, a second one is:

$$\frac{\varepsilon A}{g_0 - x}\frac{dv}{dt} + \frac{\varepsilon A}{(g_0 - x)^2}v\frac{dx}{dt} = \frac{dq}{dt} = i \tag{2-5}$$

They are suitable to describe the imposed voltage and imposed charge circuits behavior, nevertheless their derivation does not appear often intuitive, particularly in the second case, although second equation (2-5) actually corresponds to the time derivative of the constitutive relation (2-4) between charge and voltage.

The literature demonstrated that although D'Alembert's and Kirchoff's approaches are well known, straightforward and fairly intuitive, they require a certain practice to the user in balancing the applied actions: forces, current and voltage. Moreover, in case of multiple degrees of freedom and in presence of power amplifiers and related electronic devices, above limits are more evident. Therefore since 60's Crandall-Karnopp (1968) proposed to deal with the formulation of the equations of motion of electromechanical systems, by a procedure which usually looks more systematic, to describe the electromechanical energy conversion, and easier to be automated, by implementing in numerical tools. The latter is a generalization of the Lagrange's approach to the electromechanical coupled analysis.

## 2.4 Lagrange's generalized formulation

The same result obtained in previous section 2.3 can be achieved by the Lagrange's method, based on the computation of energetic contributions, used to write in Lagrangian coordinates the dynamic equilibrium equations as follows:

$$\frac{d}{dt}\left(\frac{\partial L(\xi,\dot{\xi},t)}{\partial \dot{\xi}_i}\right) - \frac{\partial L(\xi,\dot{\xi},t)}{\partial \xi_i} = Q_i \qquad i = 1,\dots,n \tag{2-6}$$

In particular, a number $n$ of independent and so-called Lagrangian coordinates $\xi_i$, strictly sufficient to describe the system generalized motion, through the so-called Lagrangian function $L$, is selected, i.e., in present case, displacement $x$ and either voltage $v$ or charge $q$. Generalized forces appear at the right-hand of the whole equations, as $Q_i$. Equations include derivatives of $L$ with respect to time $t$, to coordinates $\xi_i$, and to their generalized velocities $\dot{\xi} = d\xi/dt$.

For a pure mechanical system Lagrangian coordinates are usually: displacement $x$ and momentum $p$, or rotation $\theta$ and angular momentum $l$. The Lagrangian function $L$ is defined as:

$$L = T - U \tag{2-7}$$

being $T$ the kinetic energy and $U$ the elastic potential energy of the system. In case of pure electrical circuit selections are distinguished for two classical formulation, called "loop" and "node", according to the network analysis criteria.

Loop formulation, applied to imposed voltage configuration, introduces charge $q$ as Lagrangian coordinate. Lagrangian function is defined as:

$$L = W_m^* - W_e \tag{2-8}$$

Subscripts $m$ and $e$ in (2-8) indicate magnetic and electric contributions respectively, while symbol * distinguishes so-called co-energy of the system from the energy, represented by the capital letter $W$. Definitions of the whole two concepts shall be shortly herewith introduced.

Node formulation, typical for imposed charge configuration, is based on magnetic flux coordinate $\lambda$, which can be correlated to voltage $v$ by the following relation:

$$v = \mathrm{d}\,\lambda/\mathrm{d}\,t \tag{2-9}$$

In this case $L$ is:

$$L = W_e^* - W_m \tag{2-10}$$

**Remark 1: Selection of the Lagrangian coordinates**

In previous description the selection of charge and magnetic flux as Lagrangian coordinates may look peculiar, being people prone to use voltage with charge. The above choice is motivated by applicable electromechanical analogy, which allows transforming electrical circuit into an equivalent mechanism and vice versa. Assuming as relevant coordinates: displacement and momentum for mechanics and charge and flux for electrical systems, analogy holds as shown in figure 3. In particular, looking at the left side of sketch and starting from the lower and left corner it can be appreciated that formal analogy applies between inductor and spring: as well as magnetic flux ($\lambda$) times the inverse of inductance (*1/L*) gives current (*i*), function of charge $q$, in elastic springs displacement ($x$) times stiffness ($k$) gives the applied force (*f*), function of momentum ($p$). Other analogies can be appreciated by following paths drawn, by comparing resistor and viscous damper, up to capacitor and inertia. Described paths bring the Reader from one Lagrangian coordinate to the other, from flux to charge on electric domain, from displacement to momentum on mechanical side (see Crandall,1968).

It can be noticed that above analogy, quite popular in the literature of mechatronic systems, may be replaced by a different interpretation shown in figure 4: analogy is formally different from the previous one, although constitutive laws of mechanics and electromagnetism are the same. In particular, inductance is there analogue to inertia and capacitor to spring.

**Remark 2: Definition of energy and co-energy**

Above sections highlighted, in case of pure electrical system, the presence into the Lagrangian function of energies and complementary energies or "co-energies" (superscript *). Formally their definitions are:

$$W_e = \int_0^q v\mathrm{d}q \qquad W_m = \int_0^\lambda i\mathrm{d}\lambda \qquad W_e^* = \int_0^v q\mathrm{d}v \qquad W_m^* = \int_0^i \lambda\mathrm{d}i \tag{2-11}$$

In practice a distinction between energy and co-energy contributions can be evidenced by considering the characteristic curve of each circuit component. In a more general case of nonlinear characteristic curves, depicted in figure 5, above definitions are interpreted as areas on two diagrams, related to electric and magnetic actuations, representing: in a first case generalized force voltage versus charge coordinate, and in the second one current versus magnetic flux.

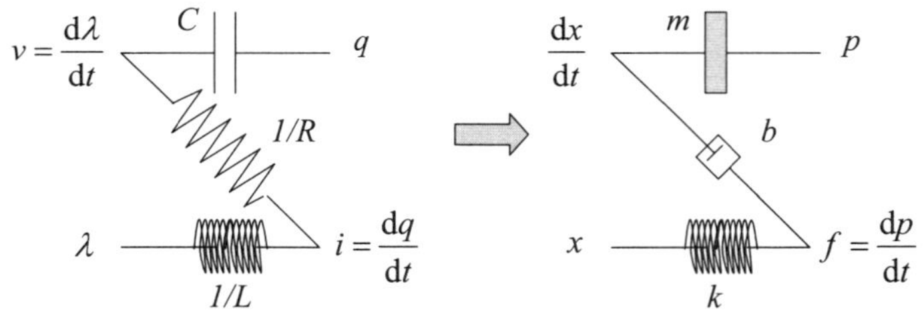


**Figure 3.** Electromechanical analogy.

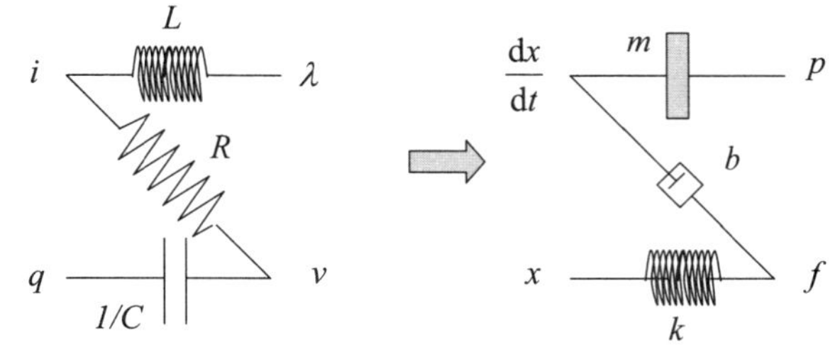


**Figure 4.** A second interpretation proposed in the literature for the electromechanical analogy.

The sum of both contributions give the overall energy storage:

$$qv = W_e^* + W_e \qquad \lambda i = W_m^* + W_m \tag{2-12}$$

In case of an electromechanical coupled system above definitions are applied to write the Lagrangian function, for the loop formulation, i.e. for imposed voltage:

$$L = T - U + W_m^* - W_e \tag{2-13}$$

and for the node formulation, i.e. for imposed charge:

$$L = T - U + W_e^* - W_m \tag{2-14}$$

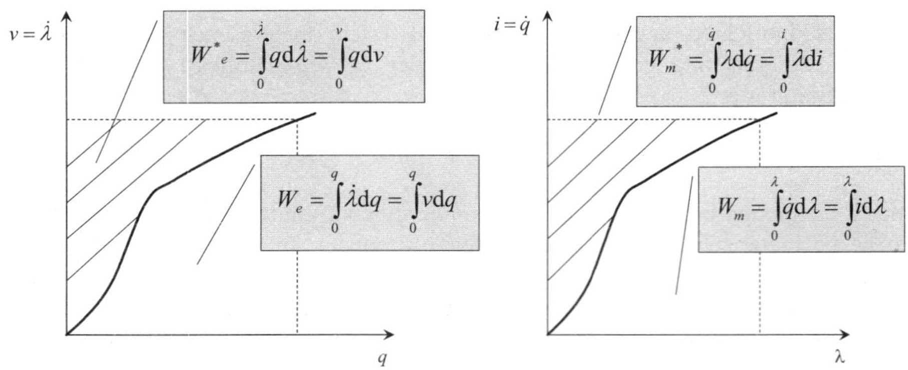


**Figure 5.** Energy and co-energy definitions.

## 2.5 Application to the electrostatic microactuator

The formulation of the equations of motion for the test case selected (see figures 1,2) can be performed by Lagrange's approach. For this purpose are useful the definition of capacitance and the constitutive law of the capacitor, respectively:

$$C = \frac{\varepsilon A}{(g_0 - x)} \qquad q = \frac{\varepsilon A}{(g_0 - x)} v \tag{2-15}$$

So-called node equation, for imposed charge, is derived as first. Selected Lagrangian coordinates are displacement $x$ and magnetic flux $\lambda$, being charge a forcing term at right hand of the equation. To write Lagrangian function $L$, energy and co-energy contributions are computed. Mechanical ones, i.e., kinetic and elastic potential, are respectively:

$$T = \frac{1}{2} m\dot{x}^2 \qquad U = \frac{1}{2} kx^2 \tag{2-16}$$

while electromagnetic ones, including electric co-energy and magnetic energy are:

$$W_e^* = \frac{1}{2} Cv^2 = \frac{1}{2} C\dot{\lambda}^2 = \frac{1}{2} \frac{\varepsilon A}{(g_0 - x)} v^2 \qquad W_m = 0 \tag{2-17}$$

Lagrangian function for node equation holds:

$$L = T - U + W_e^* - W_m \tag{2-18}$$

Since coordinates are two, a couple of equations of motion is written. Degrees of freedom are coupled, therefore only roughly speaking they can be indicated as mainly mechanical and electrical.

In particular a first, so-called mechanical equation holds:

$$\frac{d}{dt}\left(\frac{\partial L}{\partial \dot{x}}\right)-\frac{\partial L}{\partial x}=Q_m$$

$$\frac{d}{dt}\left(\frac{\partial L}{\partial \dot{x}}\right)=m\ddot{x} \qquad \frac{\partial L}{\partial x}=-kx+\frac{1}{2}\frac{\varepsilon A}{(g_0-x)^2}\dot{\lambda}^2 \tag{2-19}$$

$$m\ddot{x}+kx-\frac{\varepsilon A}{2(g_0-x)^2}\dot{\lambda}^2=F$$

Electromechanical action may appear either at left or right hand, but formally it has to be expressed in Lagrangian coordinate at left side (magnetic flux term) and in forcing term (charge) at right side:

$$m\ddot{x}+kx-\frac{\varepsilon A}{2(g_0-x)^2}\dot{\lambda}^2=F \Rightarrow m\ddot{x}+kx=F+\frac{q^2}{2\varepsilon A} \tag{2-20}$$

The latter comment is ineffective from the theoretical point of view, but it shall be useful in introducing numerical methods for solution.

A similar rationale is followed to write the second, mainly electrical, equation:

$$\frac{d}{dt}\left(\frac{\partial L}{\partial \dot{\lambda}}\right)-\frac{\partial L}{\partial \lambda}=Q_e$$

$$\frac{d}{dt}\left(\frac{\partial L}{\partial \dot{\lambda}}\right)=\frac{\varepsilon A\ddot{\lambda}}{(g_0-x)}+\varepsilon A\frac{\dot{\lambda}\,\dot{x}}{(g_0-x)^2} \qquad \frac{\partial L}{\partial \lambda}=0 \tag{2-21}$$

$$\varepsilon A\frac{\ddot{\lambda}}{(g_0-x)}+\varepsilon A\frac{\dot{\lambda}\dot{x}}{(g_0-x)^2}=i$$

Very often engineers are used thinking in terms of voltage: although following formalism cannot be accepted as canonical in structural mechatronics, it allows solving by substitution the set of two equations by computing voltage by means of the circuit equation:

$$\begin{gathered} m\ddot{x}+kx-\frac{\varepsilon A}{2(g_0-x)^2}v^2=F \Rightarrow m\ddot{x}+kx=F+\frac{\varepsilon A}{2(g_0-x)^2}v^2 \\ \varepsilon A\frac{\dot{v}}{(g_0-x)}+\varepsilon A\frac{v\dot{x}}{(g_0-x)^2}=i \end{gathered} \tag{2-22}$$

In case of imposed voltage, i.e. of loop formulation in displacement $x$ and charge $q$, procedure above introduced is developed by starting from energies and co-energies:

$$T=\frac{1}{2}m\dot{x}^2 \quad U=\frac{1}{2}kx^2 \quad W_e=\frac{1}{2}\frac{q^2}{C}=\frac{1}{2}\frac{q^2}{\varepsilon A}(g_0-x) \quad W_m^*=0 \tag{2-23}$$

to write Lagrangian function $L$:

$$L = T - U + W_m^* - W_e \tag{2-24}$$

and two equations of motion, i.e. mechanical:

$$\frac{d}{dt}\left(\frac{\partial L}{\partial \dot{x}}\right) - \frac{\partial L}{\partial x} = Q_m \qquad \frac{d}{dt}\left(\frac{\partial L}{\partial \dot{x}}\right) = m\ddot{x} \qquad \frac{\partial L}{\partial x} = -kx + \frac{1}{2}\frac{q^2}{\varepsilon A}$$

$$m\ddot{x} + kx - \frac{q^2}{2\varepsilon A} = F \Rightarrow m\ddot{x} + kx = F + \frac{\varepsilon A v^2}{2(g_0 - x)} \tag{2-25}$$

and electrical, or circuit, equation:

$$\frac{d}{dt}\left(\frac{\partial L}{\partial \dot{q}}\right) - \frac{\partial L}{\partial q} = Q_e = v \qquad \frac{d}{dt}\left(\frac{\partial L}{\partial \dot{q}}\right) = 0 \qquad \frac{\partial L}{\partial q} = -\frac{(g_0 - x)}{\varepsilon A}q$$

$$\frac{(g_0 - x)}{\varepsilon A}q = v \tag{2-26}$$

**Remark 3: Matrix form and electromechanical coupling analysis**

Above equations can be ordered in matrix form. Node equations, for instance, hold as follows:

$$\begin{bmatrix} m & 0 \\ 0 & \dfrac{\varepsilon A}{(g_0 - x)} \end{bmatrix}\begin{Bmatrix} \ddot{x} \\ \ddot{\lambda} \end{Bmatrix} + \begin{bmatrix} 0 & -\dfrac{\varepsilon A \dot{\lambda}}{2(g_0 - x)^2} \\ \dfrac{\varepsilon A \dot{\lambda}}{(g_0 - x)^2} & 0 \end{bmatrix}\begin{Bmatrix} \dot{x} \\ \dot{\lambda} \end{Bmatrix} + \begin{bmatrix} k & 0 \\ 0 & 0 \end{bmatrix}\begin{Bmatrix} x \\ \lambda \end{Bmatrix} = \begin{Bmatrix} F \\ i \end{Bmatrix} \tag{2-27}$$

Equations are interdependent, since generalized damping matrix includes non null elements out of the main diagonal. Moreover these terms are themselves nonlinear in the coupled coordinates, in particular a quadratic dependence on displacement appears. In this case a diagonal element of the generalized mass matrix is also dependent on displacement, by causing an additional coupling effect. Nevertheless the location of coupled or mixed elements strictly depends on the coordinates selection. Here above a canonical set of Lagrangian coordinates was used.

In the literature a different coupling is found, if voltage is directly assumed as electric coordinate:

$$\begin{bmatrix} m & 0 \\ 0 & 0 \end{bmatrix}\begin{Bmatrix} \ddot{x} \\ \ddot{v} \end{Bmatrix} + \begin{bmatrix} 0 & 0 \\ \dfrac{\varepsilon A v}{(g_0 - x)^2} & \dfrac{\varepsilon A}{(g_0 - x)} \end{bmatrix}\begin{Bmatrix} \dot{x} \\ \dot{v} \end{Bmatrix} + \begin{bmatrix} k & -\dfrac{\varepsilon A v}{2(g_0 - x)^2} \\ 0 & 0 \end{bmatrix}\begin{Bmatrix} x \\ v \end{Bmatrix} = \begin{Bmatrix} F \\ i \end{Bmatrix} \tag{2-28}$$

This interpretation, less used in mechatronics, exhibits a direct coupling in generalized damping and stiffness matrices, never in mass. Nevertheless the latter elements still are nonlinear dependent on the coordinates.

A slightly different look shows the corresponding matrix form of loop equations:

$$\begin{bmatrix} m & 0 \\ 0 & 0 \end{bmatrix}\begin{Bmatrix} \ddot{x} \\ \ddot{q} \end{Bmatrix}+\begin{bmatrix} k & -\dfrac{q}{2\varepsilon A} \\ 0 & \dfrac{(g_0-x)}{\varepsilon A} \end{bmatrix}\begin{Bmatrix} x \\ q \end{Bmatrix}=\begin{Bmatrix} F \\ v \end{Bmatrix} \tag{2-29}$$

A simple cross-coupling is appreciated in generalized stiffness matrix, elements of right column depend themselves from coordinates.

## 2.6 Dissipative microsystems

Originally Lagrange's formulation was based on the following assumptions: the electromechanical interaction be conservative, i.e., no dissipation in energy conversion occurs; constraints are holonomic, since do not vary in time; no mutual interaction is foreseen between magnetic and electric fields. Nevertheless, the whole method applies to dissipative systems too, at least when so-called Rayleigh's functions $\phi_i$ can be introduced to describe dissipation mechanism. Typical examples are mechanical viscous dampers and circuit resistors. In this case Lagrange's equations transform into:

$$\frac{d}{dt}\left(\frac{\partial L(\xi,\dot{\xi},t)}{\partial \dot{\xi}_i}\right)-\frac{\partial L(\xi,\dot{\xi},t)}{\partial \xi_i}+\frac{\partial \phi_i}{\partial \dot{\xi}_i}=Q_i \qquad i=1,\dots,n \tag{2-30}$$

Several kinds of damping are usually applied to MEMs, including structural and viscous, added to some due to fluid-structure coupling, when microdevice operates in air or fluidic environment, instead of vacuum. This topic, crucial for the current research activities, is out of the goal of these outlines, but in principle models proposed in the literature for the whole damping effects may be used in connection with Lagrange's approach for a refinement of the equations of motion of the dissipative microsystem.

## 2.7 Solution of nonlinear equations of motion of the undamped capacitive microactuator

In case of nonlinear electromechanical problem and undamped micro-electrostatic actuator with two electromechanical degrees of freedom equations to be solved are:

$$m\ddot{x}+kx-\frac{q^2}{2\varepsilon A}=F \qquad \frac{(g_0-x)}{\varepsilon A}q=v \qquad \text{(Loop)}$$

$$m\ddot{x}+kx-\frac{\varepsilon A}{2(g_0-x)^2}\dot{\lambda}^2=F \qquad \varepsilon A\frac{\ddot{\lambda}}{(g_0-x)}+\varepsilon A\frac{\dot{\lambda}\dot{x}}{(g_0-x)^2}=i \qquad \text{(Nodes)} \tag{2-31}$$

Moreover main attention will be played on the mechanical equation of motion, i.e. those on the left-side in previous set, being of primary priority for the mechanical design. As well as it happens in static behavior, operatively speaking the main problem is that actuation force is nonlinear dependent on both electrical and mechanical coordinates.

The expressions for the two mentioned formulations are:

$$F_e = \frac{1}{2}\frac{\varepsilon A v^2}{(g_0 - x)^2} \qquad\qquad F_e = \frac{1}{2}\frac{q^2}{\varepsilon A} \tag{2-32}$$

where $v, q, x$ appear as quadratic terms, $v$ and $q$ can be either constant or dependent on time.

Being impossible to solve analytically the whole nonlinear equations, numerical integration is applied. Known algorithms proposed for mechanical systems are distinguished in explicit methods, like the Central Finite Differences, and implicit methods, like Houbolt, Wilson-$\theta$ and Newmark. Implicit are often preferred, since they benefit of an unconditioned numerical stability, at least for linear analysis. More details on this topic can be found in (Bathe,1996; Cook, 1989). In particular Newmark method demonstrated to be very efficient in solving this kind of problems, and it is herewith applied. No mechanical excitation is assumed to be applied.

Equations to be solved are, for the two formulations:

$$m\ddot{x} + kx = \frac{\varepsilon A}{2(g_0 - x)^2}v^2 \qquad\qquad m\ddot{x} + kx = \frac{q^2}{2\varepsilon A} \tag{2-33}$$

It can be appreciated that two configurations can be dealt with, in solving above equations. If dynamic behavior is due to a mechanical excitation in presence of applied constant voltage or charge, problem consists simply of a mechanical structure whose stiffness is calibrated by the electric field. By converse, if dynamics is induced by a control voltage or charge, variable in time, it consists of the dynamic and control of an actuator. Although in the literature the two above configurations are both found, the first one seems useless for practical applications. The most attractive is the second one, since it plays the role of mechatronic actuator (for microswitches and microresonators).

## 2.8 Numerical results for the nonlinear and undamped capacitive microactuator

If the imposed voltage condition is firstly considered the corresponding equation is:

$$m\ddot{x} + kx = \frac{\varepsilon A}{2(g_0 - x)^2}v^2 \tag{2-34}$$

Numerical inputs chosen for the following numerical investigations are: $m$=9.437·10$^{-4}$ μg; $k$= 4.89 N/m; $A$=810 μm$^2$; $g_0$=2.2μm; $\varepsilon$=8.854·10$^{-12}$. Initial conditions are set as $x(0)=0$, $dx/dt(0)=0$. Applied voltage is a simple sinusoidal excitation $v$=23±0.23 V, being the pull-in voltage of the whole system $v_{PULL\text{-}IN}$=46.4 V.

The natural frequency of the pure mechanical system ($v$=0) is 362.38 kHz; while, if the decrease of stiffness due to bias voltage, is considered, frequency becomes 348.70 kHz. The latter actually corresponds to the electromechanical resonance of the microsystem, if the resonance frequency is computed by following a linearized approach, as it will be proposed in next sections. At this level these two values are useful only in terms of locating the frequency of the excitation above or below the apparent resonances of the microsystem.

A first test can be performed by exciting the whole microactuator at 346 kHz, i.e. at a frequency slightly below both the resonances. Dynamic response in terms of displacement looks like in figure 6. Although scales are poorly readable the main information to be appreciated is that displacement varies dynamically in time, by achieving a certain maximum, but always keeping finite value, i.e. amplitude does not grow infinitely, like in a linear resonant system. This behavior applies, even if no damping is applied. Moreover, the monitored signal shows a beat, which is due to the value of the exciting frequency, fairly close to a resonance of the structure. The maximum displacement corresponds to 0.65 μm, while an offset in the wave monitored towards the top of the diagram is due to the static displacement achieved because of the bias voltage, corresponding to 0.11 μm. A similar response can be found by exciting the whole microsystem at 350 kHz. In this case the value is slightly above the computed linearized resonance frequency. Figure 7 shows that amplitude is still limited, at a maximum value of 0.25 μm, less than in previous case. The two responses detected are due to the excitation frequency values: in both cases they are so closed to that of a resonance of the system, to generate a beat. The maximum value of the first is higher, since it looks almost superimposed to a resonance.

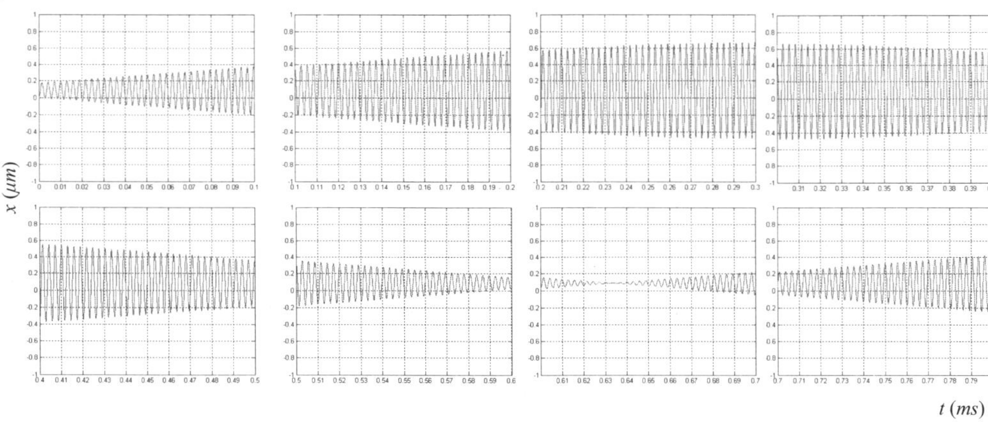


**Figure 6.** Displacement versus time, for a frequency below mechanical and electromechanical resonances (signal flows in time from left-side to right-side, continuously from first row to second one). Ranges are 0 – 0.8 ms and -1 – 1 μm.

Relevant result is that even for undamped system, amplitude is limited. An explanation of this effect can be drawn by comparing linear and nonlinear models. Linearization procedure shall be consequently presented in next section. Nevertheless, it can be already appreciated that nonlinear model and solution make available only the dynamic response in time, never in frequency domain, since superposition of effects does not applies. In particular for nonlinear analysis mechanical and electromechanical resonance is a vanishing concept, due to the particular path assumed by the curve displacement versus frequency, leading either to softening or hardening effects, dependent on the value of displacement achieved. Finally considerations about dynamic instabilities are hard to be drawn, while the literature proposed several interpretations to verify the dynamic stability about a selected equilibrium condition, based on linearized models.

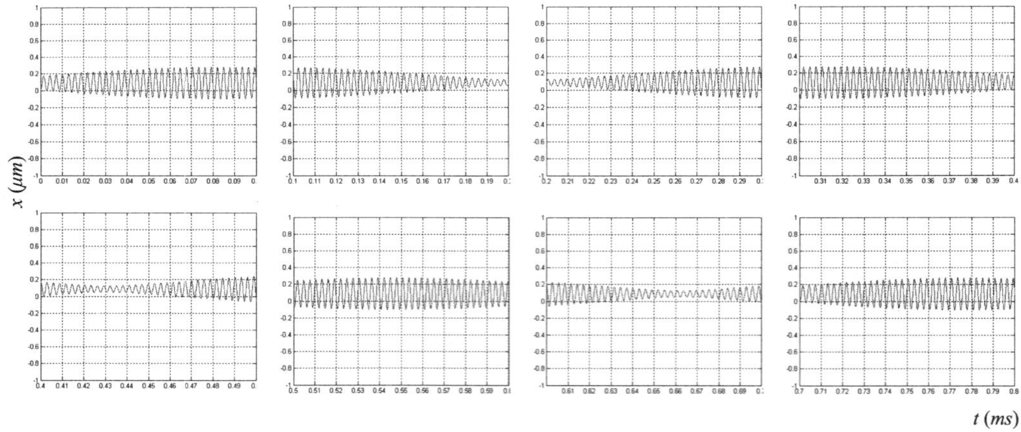

**Figure 7.** Displacement versus time, for a frequency above resonance (signal flowing is the same of above sketch). Ranges are 0 – 0.8 ms and -1 – 1 μm

## 2.9 Linearization of the equations of motion

This procedure is often applied to find an analytical solution for the equations of motion, by usually reducing the computational time for solving, or to deal with stability analysis about a selected equilibrium configuration and with frequency domain analysis.

In present case linearization applies to force $F_e$: dynamic response is assumed to be limited in amplitude about a selected electromechanical equilibrium condition, expressed by a set of fixed values of the Lagrangian coordinates, for instance:

$$x_0, \dot{x}_0 = 0, \lambda_0 = 0, \dot{\lambda}_0 = v_0 \tag{2-35}$$

Electromechanical force is linearized as follows:

$$\begin{aligned} F_e &= \frac{v^2 \varepsilon A}{2(g_0 - x)^2} = h(x,v) \cong h_0(x_0, v_0) + \left.\frac{\partial h}{\partial x}\right|_0 (x - x_0) + \left.\frac{\partial h}{\partial v}\right|_0 (v - v_0) \\ F_e &\cong \frac{v_0^2 \varepsilon A}{2(g_0 - x_0)^2} + \frac{v_0^2 \varepsilon A}{(g_0 - x_0)^3}(x - x_0) + \frac{v_0 \varepsilon A}{(g_0 - x_0)^2}(v - v_0) \end{aligned} \tag{2-36}$$

Sometimes in the literature voltage is constant and previous expression degenerates into:

$$F_e \cong \frac{v_0^2 \varepsilon A}{2(g_0 - x_0)^2} + \frac{v_0^2 \varepsilon A}{(g_0 - x_0)^3}(x - x_0) \tag{2-37}$$

**Remark 6: Electrical negative stiffness**

Above linearized expression includes three terms: a constant contribution, a dependence on relative displacement and one on the voltage increment. As it is usual for actively suspended systems, the whole formulation can be introduced into equation of motion, by splitting terms in: an additional and negative stiffness, on left hand, and a forcing term, on right hand:

$$m\ddot{x}(t)+\left(k-\frac{v_0^2\varepsilon A}{(g_0-x_0)^3}\right)x(t)=\frac{v_0\varepsilon A}{(g_0-x_0)^2}\left(v(t)-v_0\left(\frac{1}{2}+\frac{x_0}{(g_0-x_0)}\right)\right) \tag{2-38}$$

The whole operation allows appreciating that electric field applies two effects on the controlled structure: one always welcome, since forcing term is used to govern dynamic response of the microsystem, while electrical negative stiffness $k_e$ is unwanted, since it decreases the overall stiffness of the microactuator, consisting of *(k-k$_e$)*, and it is unstabilizing:

$$m\ddot{x}+(k-k_e)x=F_e \tag{2-39}$$

To complete the procedure, linearization is applied to the circuit equation:

$$\varepsilon A\frac{\ddot{\lambda}}{(g_0-x)}+\varepsilon A\frac{\dot{\lambda}\dot{x}}{(g_0-x)^2}=i$$

$$\frac{\varepsilon A}{(g_0-x_0)}\left[\dot{v}_0+\frac{v_0\dot{x}_0}{(g_0-x_0)}\right]+\left[\frac{\varepsilon A\dot{v}_0}{(g_0-x_0)^2}+\frac{2\varepsilon Av_0\dot{x}_0}{(g_0-x_0)^3}\right](x-x_0)+ \tag{2-40}$$

$$+\frac{\varepsilon Av_0}{(g_0-x_0)^2}(\dot{x}-\dot{x}_0)+\frac{\varepsilon A\dot{x}_0}{(g_0-x_0)^2}(v-v_0)+\frac{\varepsilon A}{(g_0-x_0)}(\dot{v}-\dot{v}_0)=i$$

The linearized set of coupled equations of the dynamic equilibrium is therefore:

$$\begin{bmatrix} m & 0 \\ 0 & \dfrac{\varepsilon A}{(g_0-x_0)}\end{bmatrix}\begin{Bmatrix}\ddot{x}\\ \ddot{\lambda}\end{Bmatrix}+\begin{bmatrix} 0 & -\dfrac{\varepsilon Av_0}{(g_0-x_0)^2} \\ \dfrac{\varepsilon Av_0}{(g_0-x_0)^2} & \dfrac{\varepsilon A\dot{x}_0}{(g_0-x_0)^2}\end{bmatrix}\begin{Bmatrix}\dot{x}\\ \dot{\lambda}\end{Bmatrix}+$$

$$+\begin{bmatrix} k-\dfrac{\varepsilon Av_0^2}{(g_0-x_0)^3} & 0 \\ \dfrac{\varepsilon A\dot{v}_0}{(g_0-x_0)^2}+2\dfrac{\varepsilon Av_0\dot{x}_0}{(g_0-x_0)^3} & 0\end{bmatrix}\begin{Bmatrix}x\\ \lambda\end{Bmatrix}= \tag{2-41}$$

$$=\begin{Bmatrix} -\dfrac{1}{2}\dfrac{\varepsilon Av_0^2}{(g_0-x_0)^2}-\dfrac{\varepsilon Av_0^2x_0}{(g_0-x_0)^3} \\ i-\dfrac{\varepsilon Av_0\dot{x}_0}{(g_0-x_0)^2}+\dfrac{\varepsilon A\dot{v}_0x_0}{(g_0-x_0)^2}+\dfrac{2\varepsilon Av_0x_0\dot{x}_0}{(g_0-x_0)^3}+\dfrac{2\varepsilon Av_0\dot{x}_0}{(g_0-x_0)^2}\end{Bmatrix}$$

It still holds the correspondence: $\dot{\lambda}=v;\ddot{\lambda}=\dot{v}$. If the following assumption applies $x_0\neq 0;\dot{x}_0=0;\lambda_0=0,\dot{\lambda}_0=v_0\neq 0;\ddot{\lambda}_0=\dot{v}_0=0$, the latter can be presented as:

$$\begin{bmatrix} m & 0 \\ 0 & \dfrac{\varepsilon A}{(g_0 - x_0)} \end{bmatrix} \begin{Bmatrix} \ddot{x} \\ \ddot{\lambda} \end{Bmatrix} + \begin{bmatrix} 0 & -\dfrac{\varepsilon A v_0}{(g_0 - x_0)^2} \\ \dfrac{\varepsilon A v_0}{(g_0 - x_0)^2} & 0 \end{bmatrix} \begin{Bmatrix} \dot{x} \\ \dot{\lambda} \end{Bmatrix} +$$
$$\begin{bmatrix} k - \dfrac{\varepsilon A v_0^2}{(g_0 - x_0)^3} & 0 \\ 0 & 0 \end{bmatrix} \begin{Bmatrix} x \\ \lambda \end{Bmatrix} = \begin{Bmatrix} -\dfrac{1}{2}\dfrac{\varepsilon A v_0^2}{(g_0 - x_0)^2} - \dfrac{\varepsilon A v_0^2 x_0}{(g_0 - x_0)^3} \\ i \end{Bmatrix} \tag{2-42}$$

It can be appreciated that all the above matrices exhibit now the classic symmetric or skew-symmetric structure of linear coupled controlled systems. The generalized damping matrix is skew-symmetric, since no damping is included, only the circulatory part is expressed (Inman, 2001). In case of loop formulation a dual expression can be easily drawn.

## 2.10 Solution of the linearized equations of motion

Being focused the mechanical part of the above set of equations:

$$m\ddot{x}(t) + \left(k - \frac{v_0^2 \varepsilon A}{(g_0 - x_0)^3}\right) x(t) = \frac{v_0 \varepsilon A}{(g_0 - x_0)^2}\left(v(t) - v_0\left(\frac{1}{2} + \frac{x_0}{(g_0 - x_0)}\right)\right) \tag{2-43}$$

solution can be drawn preliminarily for the free motion:

$$m\ddot{x}(t) + \left(k - \frac{v_0^2 \varepsilon A}{(g_0 - x_0)^3}\right) x(t) = 0 \tag{2-44}$$

This case corresponds in practice to a tunable spring, whose stiffness is calibrated by means of a constant applied voltage $v_0$. Linearization allows computing the frequency of resonance of the electromechanical coupled system:

$$f_r = \frac{\omega_r}{2\pi} = \frac{1}{2\pi}\sqrt{\frac{k - \dfrac{\varepsilon A {v_0}^2}{(g_0 - x_0)^3}}{m}} = \frac{1}{2\pi}\sqrt{\frac{k - k_e}{m}} \tag{2-45}$$

It is an analytical expression, constant for a given voltage $v_0$, decreasing with voltage and it is lower than the pure mechanical resonance:

$$f_0 = \frac{\omega_0}{2\pi} = \frac{1}{2\pi}\sqrt{\frac{k}{m}} \tag{2-46}$$

Solution in terms of displacement is:

$$x(t) = X_1 \sin(\omega_r t) + X_2 \cos(\omega_r t) \tag{2-47}$$

where amplitudes $X_1$ and $X_2$ are computed according to the boundary conditions.

Forced vibration can be predicted by solving the linearized equation:

$$m\ddot{x}(t)+\left(k-\frac{v_0^2\varepsilon A}{(g_0-x_0)^3}\right)x(t)=\frac{v_0\varepsilon A}{(g_0-x_0)^2}\left(v(t)-v_0\left(\frac{1}{2}+\frac{x_0}{(g_0-x_0)}\right)\right) \tag{2-48}$$

In particular, a possible expression for the imposed variable voltage is:

$$v(t)=v_0+\Delta v(t)=v_0+\Delta V\sin(\omega_V t) \tag{2-49}$$

leading to:

$$m\ddot{x}(t)+\left(k-\frac{v_0^2\varepsilon A}{(g_0-x_0)^3}\right)x(t)=\frac{v_0\varepsilon A}{(g_0-x_0)^2}\left(\Delta v(t)+v_0\left(\frac{1}{2}-\frac{x_0}{(g_0-x_0)}\right)\right) \tag{2-50}$$

where: $v_0$ is the bias voltage (or static component), $\Delta v(t)$ the dynamic component of voltage (variable), $\Delta V$ is the amplitude of the dynamic component, $\omega_V$ corresponds to the excitation frequency.

Solution can be written as sum of the previous free motion and of the current forced term:

$$x(t)=X_1\sin(\omega_r t)+X_2\cos(\omega_r t)+\frac{\psi_1}{\omega_r^2-\omega_V^2}\sin(\omega_V t)+\frac{\psi_2}{\omega_r^2}$$

$$\psi_1=\frac{v_0\varepsilon A}{m(g_0-x_0)^2}\Delta V \qquad \psi_2=\frac{v_0^2\varepsilon A}{2m(g_0-x_0)^2} \tag{2-51}$$

Symbols $\omega_V, \omega_r$ are related to the voltage command and the system response, respectively.

## 2.11 Numerical results for the linearized and undamped capacitive microactuator

The equation solved is:

$$m\ddot{x}(t)+\left(k-\frac{v_0^2\varepsilon A}{(g_0-x_0)^3}\right)x(t)=\frac{v_0\varepsilon A}{(g_0-x_0)^2}\left(v(t)-v_0\left(\frac{1}{2}+\frac{x_0}{(g_0-x_0)}\right)\right) \tag{2-52}$$

while test case parameters, as in nonlinear solution, are: $m$=9.437·10$^{-4}$ μg; $k$=4.89 N/m; $A$=810 μm$^2$; $g_0$=2.2μm; $\varepsilon$=8.854·10$^{-12}$ ; initial conditions $x(0)=0$, $dx/dt(0)=0$; excitation $v$=23±0.23 V (sinusoidal), with pull-in voltage equal to $v_{PULL\text{-}IN}$=46.4 V.

If displacement is plotted against time, at resonance, for an excitation occurring at 349 kHz, in this case it can be appreciated a continuously growing up amplitude, typical of resonance phenomenon (figure 8).

By converse, if frequency is set at 350 kHz, i.e. very closed to resonance, but slightly out of it, evolution in time looks like in figure 9. Amplitude is limited, a beat confirms that exciting and resonance frequencies are very close.

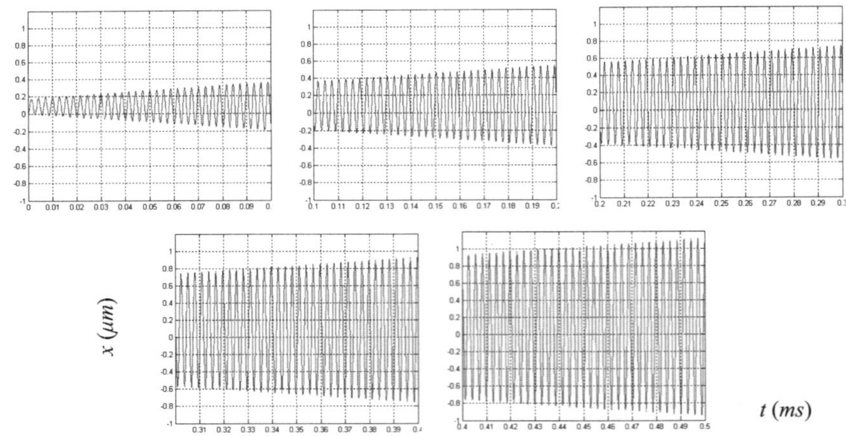


**Figure 8.** Displacement versus time, for linear solution, at resonance. Signal flows from left-side to right-side, from top to bottom. Ranges: 0 – 0.5 ms, -1 – 1 μm.

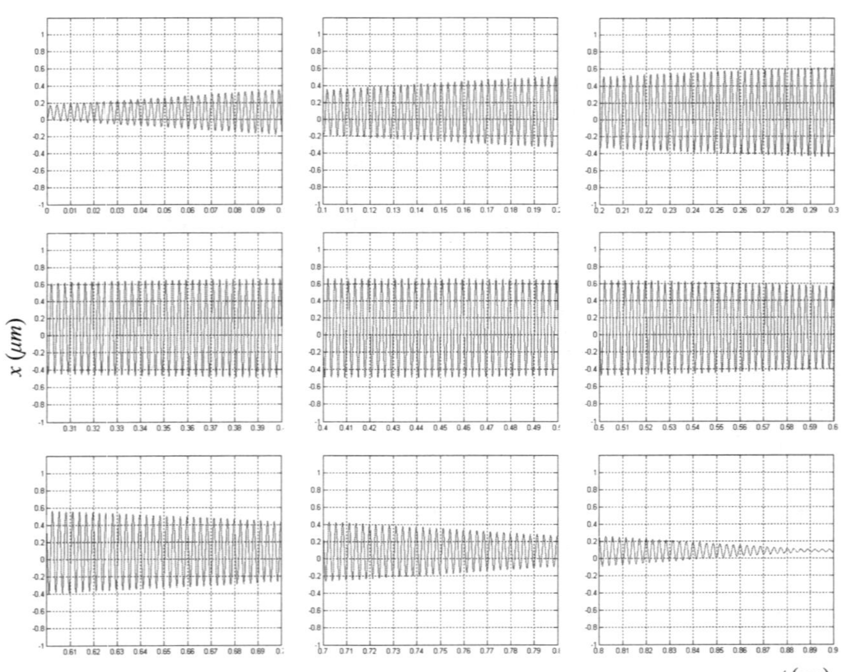


**Figure 9.** Displacement versus time, for linear solution, slightly above resonance. Ranges: 0 – 0.9 ms, -1 – 1 μm.

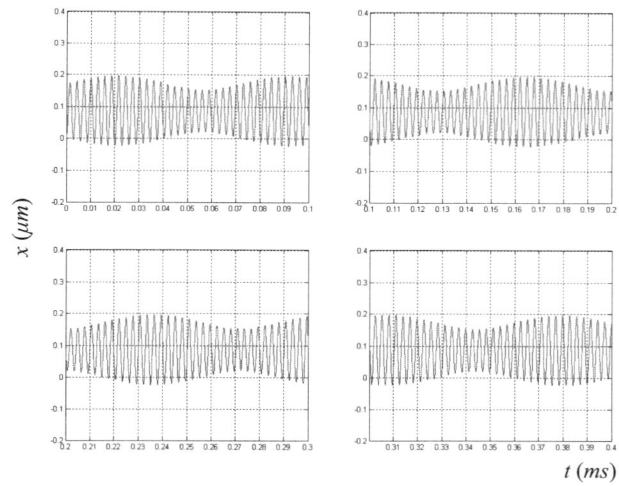

**Figure 10.** Displacement versus time, for linear solution, slightly below resonance. Ranges: 0 – 0.4 ms, -0.2 – 0.4 μm.

The expected maximum displacement is now 0.66 μm, instead of infinite, as at resonance. If a new value of exciting frequency is investigated, i.e. 335 kHz, fairly out of resonance, although not so far, result becomes like in figure 10. Beat phenomenon is still present, maximum amplitude is far less, being 0.2 μm.

# 3 Numerical and experimental characterization of dynamic behavior of micromechatronic systems

Previous sections proposed, very quickly, a systematic approach to formulate and solve electromechanically coupled problems at microscale. Numerical investigations have been focused on the solution of nonlinear and linearized models, currently discussed and tested in the literature for the design of the whole microsystems. Moreover the difference between a passively coupled microstructure, fluctuating into an electric field and an actively controlled device, under variable voltage or charge has been stressed out.

A preliminary comment to the obtained numerical results, is that linearized models allow computing a resonance of the microsystem, which is a vanishing concept for nonlinear equations. This resonance does not correspond to an actual behavior of the system, since a fast growing up of amplitude of the displacement is appreciated, on the nonlinear solution too, but is limited, never infinite, even if system is undamped. Nevertheless the detected value allows identifying at least a critical range of frequency, for which the amplitude of the dynamic response may be larger.

According to the previous aspect, linear solution allows characterizing the dynamic behavior in terms of the so-called "frequency shifting curve", often proposed in the literature. Actually, in case of nonlinear actuator characterization, the above curve is not sufficient to completely describe the investigated behavior. In next sections two effects dominant nonlinear behavior, briefly referred to as "softening" and "jumping" will be introduced to compare linear and nonlinear models predictions and to propose some criteria for an experimental characterization of the whole microsystems.

### 3.1 Frequency shifting curve computed by linearized model

The expression of the electromechanical resonance, above mentioned, is used for a characterization of the dynamic behavior of MEMs, by means of the curve frequency versus voltage, interpreted as constant value. Figure 11a shows a non-dimensional representation, for which:

$$\omega^* = \frac{\omega}{\omega_0} = \sqrt{1 - \frac{k_e}{k}}; \qquad \nu^* = \frac{v_0}{v_{\text{PULL-IN}}} \tag{3-1}$$

frequency is compared to the frequency of the mechanical system, without applied voltage, while voltage is compared to the pull-in value. Shifting phenomenon is mainly due to the bias component of voltage and is the relevant effect, affecting the dynamic behavior of microbeam under electrostatic load. In case of controlled microsystem, a second dominant effect is due to the variable part of voltage.

### 3.2 Frequency response of nonlinear model, with softening and jumping phenomena

To check the role of the variable part of voltage, a numerical test can be performed on a microactuator with: $m$=3.505·10$^{-4}$ μg; $k$=4.89 N/m; $A$=810 μm$^2$; $g_0$=2.2μm; $\varepsilon$=8.854·10$^{-12}$. The maximum displacement of the nonlinear solution of forced vibration is computed as function of the dynamic component of voltage $\Delta V$ and of the excitation frequency, for a given bias voltage, for instance equal to the 50% of pull-in. The result is depicted in figure 11b.

It can be realized that the aspect of the frequency is fairly different from that of linear case: nonlinear one exhibit a curved "backbone" instead of a straight vertical asymptote in correspondence to a fixed resonance. The effect is larger for higher values of $\Delta V$ .

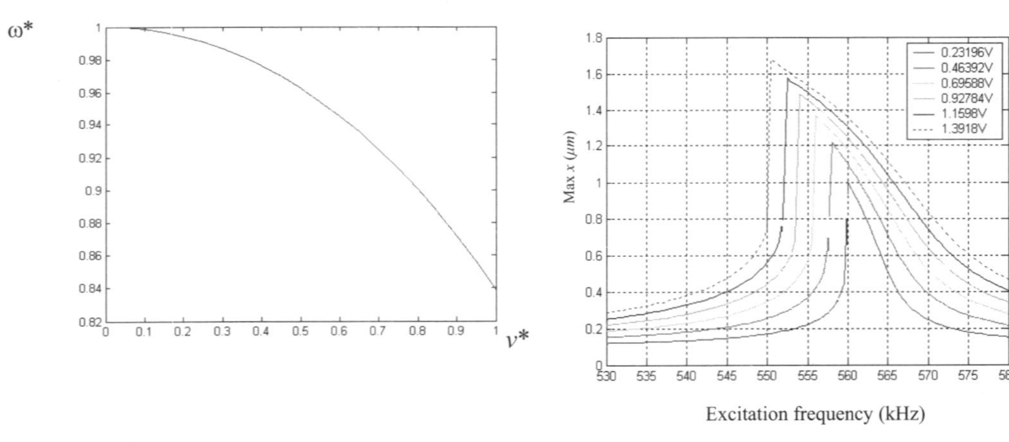


**Figure 11.** (a) So-called "frequency shifting curve" and (b) frequency curves for nonlinear electrostatic microactuator, for different values of voltage (dynamically variable contribution).

Two situations are compared in figure 12: the expected curve for linear response and the corresponding path for nonlinear behavior. The latter shows an apparent decreasing of the overall stiffness of the microsystem, for increasing displacement: it is usually referred to as softening, to be distinguished from an opposite hardening, producing a curved backbone on the right side of the diagram. This effect motivates the differences detected in previous sections, concerning responses depicted in figures 6-10.

Moreover solution of the equations of motion, in nonlinear models do not have continuous path in frequency diagram: as figure 13 shows if excitation frequency is gradually increased response amplitude can be found initially on path A-B, then solution jumps on the second plotted curve at C, by following this one up to point D. By converse if frequency is decreased from D, solution will follow curve D-C-E, then it will jump down on path B-A, at point F. The latter phenomenon is referred to as jumping phenomenon, in amplitude prediction.

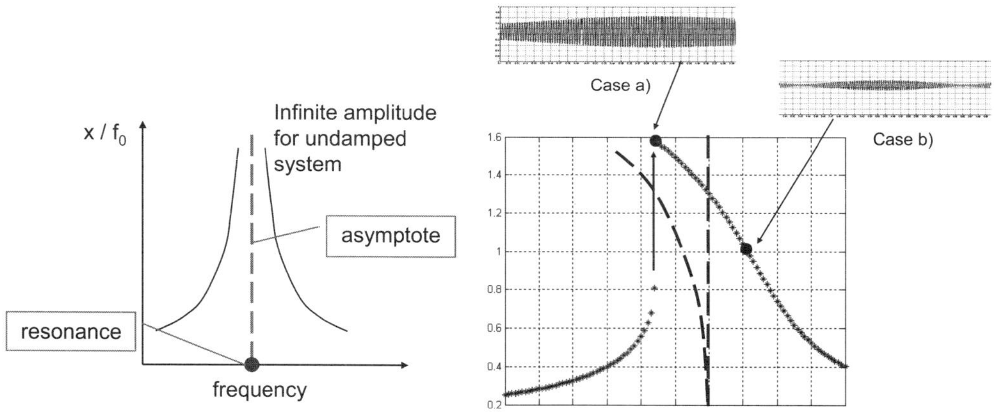


**Figure 12.** Comparison between linear and nonlinear dynamic responses in frequency domain.

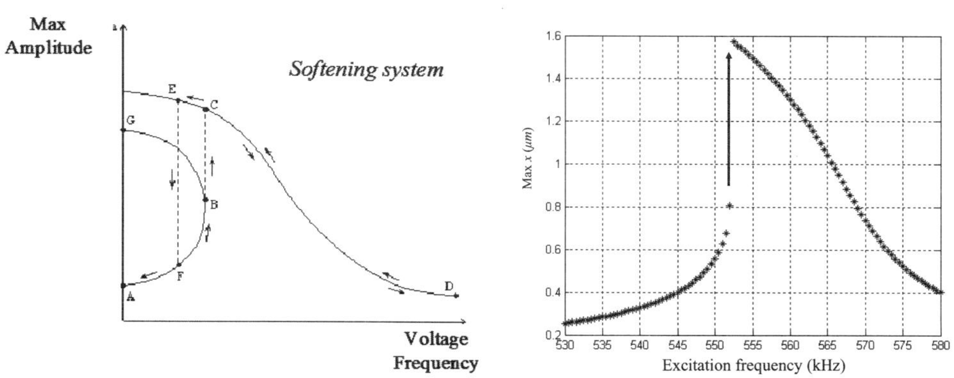


**Figure 13.** Jumping in nonlinear dynamic response.

Discontinuity in amplitude values, with respect of frequency, is effect of the combination of the whole softening and jumping phenomena, typical of nonlinear systems, never of unstable dynamic behavior.

Above discussion suggests to include in dynamic characterization more than the proposed frequency shifting effect, useful for a rough interpretation of the bias voltage on dynamic response of MEMS, a detailed analysis of nonlinear behavior, to evaluate the location in frequency of the jump on the whole diagram and the effect of softening on the amplitude of the dynamic response.

# 4 Final remarks

All comments expressed for a discrete microsystem, with two electromechanical degrees of freedom, may be extended to multi-d.o.f. systems. In particular equations of motion of the coupled electromechanical system can be derived by means of the Lagrangian approach, as well; linearization procedure is applied to make possible the above mentioned list of operations (stability analysis, frequency domain, ...); the solution of linearized equations of motion becomes an eigenproblem, in which eigenvalues give the frequency of the mechanical, electrical and electromechanical mode shapes, being eigenvectors.

A key feature of the multi d.o.f.'s modelling is the model order reduction of the d.o.f.'s to reduce the size of the problem, according to Guyan or Component mode synthesis (Genta,2000), but they are only applicable in case of linear systems, where matrices of stiffness, damping, mass are constant. As well as in case of pure mechanical system model order reduction may introduce approximation in dynamic analysis, when inertial terms of the slave d.o.f.'s are neglected.

Particularly difficult is modeling damping phenomena affecting dynamic behaviour at microscale. Several typologies of damping are present. Structural and thermoelastic damping are subject of research activity as well as squeezed-film damping. Previously proposed models are therefore coupled to fluidic effects to investigate, either through Reynold's or Navier-Stoke's equations, the triple coupling among structural, fluidic and electric behaviors. As relevant result the latter activity demonstrates that a superposition of effects, never strictly linear, modifies some general aspects of the dynamic behavior of the whole microsystem. Softening response in electromechanical coupling, for instance, may look hardening in present of fluidic interaction.

The latter aspects draw current beachmarks in modeling and understanding micromechatronic systems for a consistent design activity.

### References

Crandall, S.H., Karnopp, D.C., Pridmore-Brown D.C. (1968) *Dynamics of mechanical and electromechanical systems,* McGraw-Hill.

Woodson, H.H., Melcher, J.R. (1968) *Electromechanical dynamics,* John Wiley and Sons.

Meisel, J. (1973) *Principles of electromechanical energy conversion,* Robert Krieger Publ. Comp.

Cook, R.D. (1989) *Concepts and applications of finite element analysis,* Wiley, New York.

Meirovitch, L. (1990) *Dynamics and control of structures*, New York, John Wiley and Sons.

Junkins, J.L. (1993) *Introduction to dynamics and control of flexible structures,* AIAA series, Washington.

Bathe, K.J. (1996) *Finite element procedures*, Prentice-Hall, Englewood Cliffs.

Shi, R., Ramesh, G., Mukherjee, T. (1996) Dynamic analysis of micro-electro-mechanical system, *International Journal of Numerical Methods in Engineering*, 39, pp.4119-4139.

Ananantasuresh, G.K., Gupta, R., Senturia, S. (1996) An approach to macromodeling of MEMS for nonlinear dynamic simulation, in *DSC-Vol.59 Micromechanical Systems (MEMS),* ASME, pp.401-407.

Genta, G. (2000) *Vibration of structures and machines,* Springer Verlag.

Pauli, W. (2000) *Electrodynamics – Pauli Lectures on Physics,* vol.1, Dover.

Rizzoni,G. (2000) *Principles and applications of electrical engineering*, 3rd Edition, McGraw Hill, Int.Edition.

Inman, D. (2001) *Engineering vibration*, Prentice Hall.

Senturia, S.D. (2001) *Microsystems design*, Kluwer, Boston.

Younis, M.I., Abdel-Rahman, E.M., Nayfeh A. (2003) A reduced-order model for electrically actuated microbeam-based MEMS, *J. of Microelectromechanical Systems,* vol.12, N.5, pp.672-680.

Giurgiutiu, V.; Lyshevski, S.E. (2004) *Micromechatronics*, CRC Press, Boca Raton.

## Related articles and papers

Bona, B., Brusa, E., Canestrelli, P., Genta, G., Tonoli A. (1994) Finite Element Modeling and Experimental Validation of an Elastic Beam with Surface Bonded Piezoelectric Devices, Proc. IEEE International Conference on Robotic and Automation, San Diego, May 1994, pagg.2659-2664.

Brusa, E., Carabelli, S., Delprete, C., Tonoli A. (1996) Modeling for "self-sensing" electromechanical systems, Proc. JSME III MoViC International Conference on Motion and Vibration Control, Chiba, Japan, September 1-6 1996, vol.1, pagg.117-122.

Brusa, E., Carabelli, S., Tonoli, A.(1996) Self-sensing colocated structures with distributed piezoelectric transducers, Proc. 7th Int. Conf. on Adaptive Structures and Technologies (ICAST 96), Sept.23-25, 1996, Roma, Italia, Technomic Publ. Co .Inc., pp.84-94.

Brusa, E. (1997) *Design of electromechanical integrated structures*, PhD Thesis, Politecnico di Torino.

Brusa, E., Carabelli, S., Carraro, F., Tonoli A. (1998) Electromechanical tuning of self-sensing piezoelectric transducers, *Journal of Intelligent Material Systems and Structures*, S.I., 9 (3), pp.157-238.

Brusa, E., Carabelli, S., Tonoli A. (1998) Modelling and testing of electromechanical balanced active damped plates, Proc. SPIE 5th Annual Int. Symposium on Smart Structures and Materials, pp.879-892

Brusa, E., De Bona, F., Gugliotta, A., Somà A. (2004) Dynamics modeling of microbeams under electrostatic load, *Analog Integrated Circuits and Signal Processing*, 40(2), pp.155-164.

Munteanu, M., De Bona, F., Collenz, A., Brusa, E. (2004) Geometrical nonlinearities of electrostatically actuated microbeams, Proc. ECCOMAS 2004, Jyväskylä, Finland, ISBN 951-39-1869-6; CD-ROM and abs.vol.I, p.12.

Brusa, E., De Bona, F., Della Schiava, A., Somà, A. (2004) Model synthesis of structural dynamics for electrostatic microactuators, Proc. First ASME European Micro and Nano systems 2004, Paris, pp.87-91; ISBN 2-84813-037-7.

Munteanu, M.Gh., Brusa, E. (2005) A new FEM approach to the coupled-field analysis of electrostatic microactuators dynamics, in *Computational Methods for Coupled Problems in Science and*

*Engineering*, Ed. CIMNE, Barcelona, Spain, ISBN: 84-95999-71-4, Proc. of ECCOMAS Thematic Conference on Coupled Problems 2005, May 25-28, 2005, Santorini, Greece.

Botteon, M., Brusa, E., Munteanu, M. Gh. (2005) Coupled-field FEM nonlinear dynamics analysis of continuous microsystems by non incremental approach, Proc. IEEE, DTIP 2005; CAD, Design and Test Conference, June 1-3, 2005, Montreux, Switzerland, ISBN:2-84813-0357-1, CD-ROM

# Continuum Microstructures Loaded Electrostatically

Francesco De Bona[1], Mircea Gh. Munteanu[1, 2]

[1] Dipartimento di Ingegneria Elettrica Gestionale e Meccanica, University of Udine, Udine, Italy
[2] Cathedra Rezistenza Materialoir, Universitatea Brasov, Romania

**Abstract.** Electrostatic actuated flexible structure are frequently encountered in microsystems. The behaviour of these devices is characterized by electromechanical coupling, due to the mutual interaction between the electrostatic field and the deflection of the structure. A common case, frequently analyzed in the literature, is that of cantilever beam loaded electrostatically; in this case different analytical approaches based on a strong simplification of the elctromechanical model are available. If a more accurate analysis has to be performed, methods based on numerical techniques have to be preferred. In this case possible approaches are: lumped models, methods based on a Newton's non-linear solution scheme, sequential field coupling algorithms.

# 1 Introduction

Very often microdevices are made of flexible continuum structures whose motion is obtained by applying a distributed electrostatic load (Madou, 2002); common cases are those of cantilever and double clamped beam and plate, circular diaphragm, etc. In these cases the problem is non-linear due to the electromechanical coupling caused by the mutual interaction between the electrostatic field and the deflection of the structure.

In the following, the case of a cantilever beam loaded electrostatically will be considered; in fact this configuration is very common in microsystem design, therefore several solution approaches have been proposed in the literature.

# 2 Cantilever Microbeam under Electrostatic Loads: Analytical Solutions

Figure 1 shows the considered case: a beam of length $l$, thickness $t$ and width $w$ is clamped at one end and free at tip; an electric potential is applied between the beam and a fixed ground put at an initial distance $g_0$. The beam and the ground are made or coated (dotted line) with a conducting material.

If a plane model is considered, the force per unit length $p_e$ applied at the lower surface of the beam can be approximated by the following expression :

$$p_e = \frac{1}{2}\frac{\varepsilon w}{\left(g_0 - v(x)\right)^2} V_0^2 \tag{2-1}$$

This corresponds to mimic the electrical domain with infinite plane parallel capacitors; obviously this approximation holds only in the case of small deflections, moreover forces in the upper surface of the beam and concentrated actions at the tip are not taken into account.

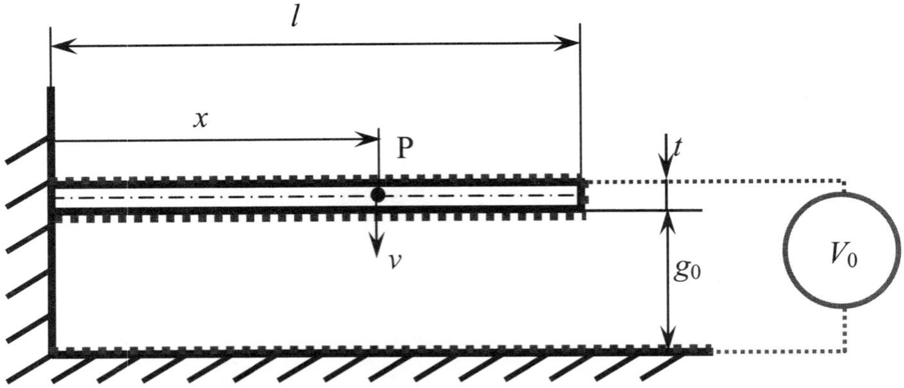


**Figure 1.** Electrostatically actuated microbeam.

An analytical solution of the previous equation is not available; according to Choi and Lovell (1997), in the case of small deflections, it could be useful to consider the Taylor expansion of the expression of the electrostatic load $p_e$:

$$p_e = \frac{w\varepsilon}{2} V_0^2 \left( \frac{1}{g_0^2} + \frac{2v}{g_0^3} + \dots \right) \tag{2-2}$$

If only the first term is considered, which corresponds to assume the electrostatic load as constant, it results:

$$p_e = \frac{w\varepsilon}{2} V_0^2 \left( \frac{1}{g_0^2} \right) \tag{2-3}$$

The governing equation of equilibrium can be expressed as::

$$\frac{d^2 v}{dx^2} = \frac{1}{EJ} p_e \frac{(l-x)^2}{2} \tag{2-4}$$

and integrating twice:

$$\frac{dv}{dx} = \frac{p_e}{EJ}\left( \frac{l^2}{2}x + \frac{x^3}{6} - l\frac{x^2}{2} \right) + C_1 \qquad v = \frac{p_e}{EJ}\left( \frac{l^2}{4}x^2 + \frac{x^4}{24} - l\frac{x^3}{6} \right) + C_1 x + C_2 \tag{2-5}$$

By imposing the boundary conditions:

$$\frac{\mathrm{d}v(0)}{\mathrm{d}x} = 0 \qquad v(0) = 0 \qquad \Rightarrow C_1 = C_2 = 0 \tag{2-6}$$

The final expression of the beam deflected shape is:

$$v = \frac{p_e}{24EJ}\left(x^4 - 4x^3 l + 6x^2 l^2\right) \;\Rightarrow v = \frac{1}{4} V_0^2 \frac{\varepsilon}{E g_0^2 t^3}\left(x^4 - 4x^3 l + 6x^2 l^2\right) \tag{2-7}$$

as :

$$p_e = \frac{w\varepsilon}{2} V_0^2 \left(\frac{1}{g_0^2}\right) \qquad J = \frac{t^3 w}{12} \tag{2-8}$$

The maximum deflection, i.e. deflection at the tip ($x=l$), is:

$$v_{\max} = \frac{3}{4}\frac{\varepsilon l^4}{E g_0^2 t^3} V_0^2 \tag{2-9}$$

If the linear term of the Taylor expansion of the electrostatic load expression is retained:

$$p_{\mathrm{e}} = \frac{w\varepsilon}{2} V_0^2 \left(\frac{1}{g_0^2} + \frac{2v}{g_0^3}\right) \tag{2-10}$$

the solution can be obtained if the following equilibrium equation is considered:

$$\frac{\mathrm{d}^4 v}{\mathrm{d}x^4} = \frac{1}{EJ} p_{\mathrm{e}} \Rightarrow \frac{\mathrm{d}^4 v}{\mathrm{d}x^4} = \frac{\varepsilon w}{2EJ} V_0^2 \left(\frac{1}{g_0^2} + \frac{2v}{g_0^3}\right) \tag{2-11}$$

That can be written:

$$\frac{\mathrm{d}^4 v}{\mathrm{d}x^4} - \mathrm{A}v = \mathrm{B} \tag{2-12}$$

where:

$$\mathrm{A} = \varepsilon w \frac{V_0^2}{EJ g_0^3}; \quad \mathrm{B} = \varepsilon w \frac{V_0^2}{2EJ g_0^2}.$$

Introducing $a = \sqrt[4]{\mathrm{A}}$ it results the differential equation:

$$\frac{\mathrm{d}^4 v}{\mathrm{d}x^4} - a^4 v = \mathrm{B} \tag{2-13}$$

whose the general solution is:

$$v = \mathrm{C}_1 \sin ax + \mathrm{C}_2 \cos ax + \mathrm{C}_3 \sinh ax + \mathrm{C}_4 \cosh ax - v_0 \tag{2-14}$$

with: $v_0 = \dfrac{\mathrm{B}}{\mathrm{A}} = \dfrac{g_0}{2}$.

By imposing the boundary conditions:

$$\begin{cases} x = 0; & v = 0 \\ x = 0; & \dfrac{\mathrm{d}v}{\mathrm{d}x} = 0 \\ x = l; & \dfrac{\mathrm{d}^2 v}{\mathrm{d}x^2} = 0 \ \ (M = 0) \\ x = l; & \dfrac{\mathrm{d}^3 v}{\mathrm{d}x^3} = 0 \ \ (T = 0) \end{cases} \tag{2-15}$$

The following linear system is obtained:

$$\begin{bmatrix} 0 & 1 & 0 & 1 \\ 1 & 0 & 1 & 0 \\ -\sin al & -\cos al & +\sinh al & +\cosh al \\ -\cos al & +\sin al & +\cosh al & +\sinh al \end{bmatrix} \begin{Bmatrix} C_1 \\ C_2 \\ C_3 \\ C_4 \end{Bmatrix} = \begin{Bmatrix} v_0 \\ 0 \\ 0 \\ 0 \end{Bmatrix} \tag{2-16}$$

with solution (MATLAB Symbolic ):

$$\begin{cases} C_1 = g_0 \dfrac{\sinh al \cos al + \sin al \cosh al}{4(1 + \cosh al \cos al)} \\ C_2 = g_0 \dfrac{1 - \sinh al \sin al + \cos al \cosh al}{4(1 + \cosh al \cos al)} \\ C_3 = -g_0 \dfrac{\sinh al \cos al + \sin al \cosh al}{4(1 + \cosh al \cos al)} \\ C_4 = g_0 \dfrac{1 + \sinh al \sin al + \cos al \cosh al}{4(1 + \cosh al \cos al)} \end{cases} \tag{2-17}$$

Pull-in instability occurs when: $1+\cosh al\cos al=0\,.$ (2-18)

It results: $al=1.8751$ $\quad w\varepsilon\dfrac{V^2}{EJg_0^3}l^4=12.36$ (2-19)

In conclusion: $V_{\text{pull-in}}=\sqrt{\dfrac{Et^3g_0^3}{\varepsilon l^4}}$ (2-20)

An alternative approach is proposed in Serraute et al. (1999) where an approximated solution of the equilibrium equation is proposed. The procedure is summarized in the following.

When a voltage $V_0$ is applied between the beam and the electrode, the beam is not deformed: the electrostatic pressure can thus initially be considered constant and equal to $p_{e0}$.

The beam will become deformed under the action of this constant pressure and the equation of the beam deformed shape can be written as:

$$v_0(x)=\frac{p_{e0}}{24EJ}\left(x^4-4x^3l+6x^2l^2\right)\Rightarrow v_0(x)=\frac{p_{e0}}{k(x)} \tag{2-21}$$

where: $\quad p_{e0}=\dfrac{1}{2}\dfrac{w\varepsilon}{g_0^{\,2}}V_0^2$

$$k(x)=\frac{24EJ}{\left(x^4-4x^3l+6x^2l^2\right)}\Rightarrow k(x)\frac{2Et^3w}{\left(x^4-4x^3l+6x^2l^2\right)} \tag{2-22}$$

The electrostatic pressure is then no longer constant: its expression depends on the $x$-abscissa:

$$p_{e1}=\frac{1}{2}\frac{w\varepsilon}{(g_0-v_0(x))^2}V_0^2=\frac{p_{e0}}{\left(1-\dfrac{v_0(x)}{g_0}\right)^2} \tag{2-23}$$

The deflection due to this non-constant pressure can be approached by:

$$v_1(x)=\frac{p_{e1}}{k(x)}\Rightarrow \quad v_1(x)=\frac{p_{e0}}{k(x)\left(1-\dfrac{v_0(x)}{g_0}\right)^2} \tag{2-24}$$

Similarly:
$$v_2(x) = \frac{p_{e2}}{k(x)} \tag{2-25}$$

as:
$$p_{e2} = \frac{1}{2}\frac{w\varepsilon}{(g_0 - v_1(x))^2}V_0^2 = \frac{p_{e0}}{\left(1 - \frac{v_1(x)}{g_0}\right)^2} \tag{2-26}$$

it follows:
$$v_2(x) = \frac{p_{e0}}{k(x)\left(1 - \frac{v_1(x)}{g_0}\right)^2} \tag{2-27}$$

In conclusion:
$$v_{n+1}(x) = \frac{p_{e(n+1)}}{k(x)} = \frac{p_{e0}}{k(x)\left(1 - \frac{v_n(x)}{g_0}\right)^2} \Rightarrow v_{n+1}(x) = \frac{v_0(x)}{\left(1 - \frac{v_n(x)}{g_0}\right)^2} \tag{2-28}$$

If the solution is convergent, $v$ is solution of:
$$v^3 - 2g_0v^2 + g_0^2v - g_0^2v_0 = 0 \tag{2-29}$$

According to Cardano's rule, if the polynomial determinant is negative:
$$D = v_0 - \frac{4}{27}g_0 < 0 \Rightarrow v_0 < \frac{4}{27}g_0 \tag{2-30}$$

three different and real solutions exit and the following trigonometric expressions hold:
$$\begin{aligned} v_1 &= \frac{2}{3}g_0\left(1 + \cos\left(\frac{1}{3}\arccos\left(\frac{27}{2}\frac{v_0}{g_0} - 1\right)\right)\right) \\ v_2 &= \frac{2}{3}g_0\left(1 + \cos\left(\frac{1}{3}\arccos\left(\frac{27}{2}\frac{v_0}{g_0} - 1\right) + \frac{2}{3}\pi\right)\right) \\ v_3 &= \frac{2}{3}g_0\left(1 + \cos\left(\frac{1}{3}\arccos\left(\frac{27}{2}\frac{v_0}{g_0} - 1\right) + \frac{4}{3}\pi\right)\right) \end{aligned} \tag{2-31}$$

The pull-in voltage is obtained putting: $D(l) = v_0 - \frac{4}{27} g_0 = 0$

$$v_0(l) = \frac{4}{27} g_0 \Rightarrow \frac{1}{2}\frac{\varepsilon}{g_0^{\ 2}} V_0^2 \frac{1}{k(l)} = \frac{4}{27} g_0 \Rightarrow \frac{1}{2}\frac{\varepsilon}{g_0^{\ 2}} V_0^2 \frac{\left(l^4 - 4l^4 + 6l^4\right)}{2Et^3} = \frac{4}{27} g_0 \quad (2\text{-}32)$$

Finally, the expression:

$$V_{\text{pull-in}} = \frac{4}{9l^2}\sqrt{\frac{g_0^3 t^3 E}{\varepsilon}} \quad (2\text{-}33)$$

is obtained occurring for a tip displacement:

$$v_{\text{PI}} = \frac{4}{27} g_0 . \quad (2\text{-}34)$$

In Figure 2 and 3 the deflected shape of the beam is shown for the three proposed methods. If the applied voltage is relatively low (see Figure 2), therefore producing a limited deflection, the agreement between the three methods is good; on the other hand, in the case of voltages quite close to pull-in (see Figure 3), the three deflection curves differ each other significantly. It has to be noticed that different values of normalized voltage $V/V_{\text{pull-in}}$ can be obtained, depending from the expression of the pull-in voltage proposed. In particular the value obtained in Serraute (1999) and that suggested in Osterberg (1997) were considered.

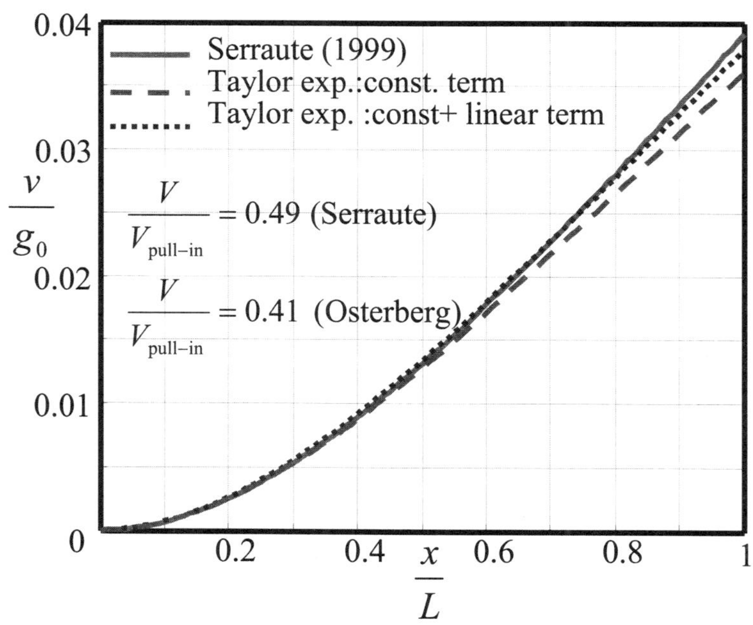


**Figure 2. Deflected microbeam shape: medium actuation voltages.**

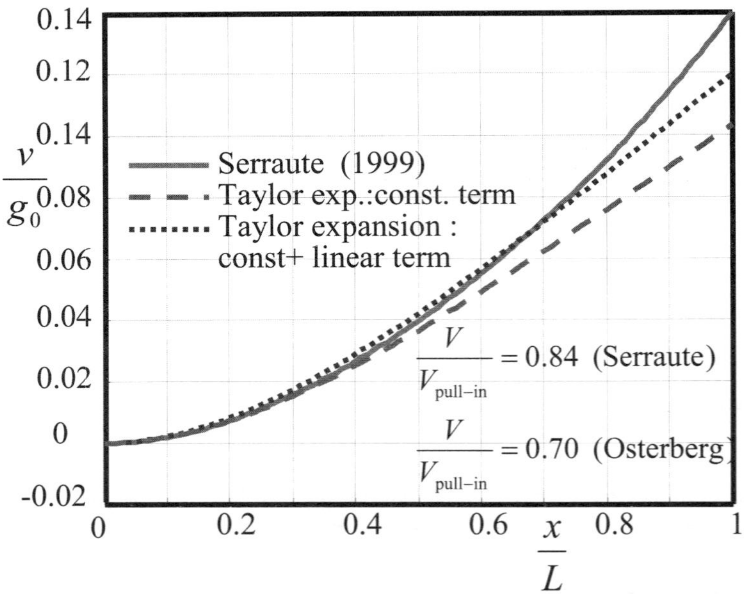


**Figure 3.** Microbeam deflected shape: high actuation voltages.

### 2.1 Pull-in Evaluation

In the case of microdevices based on electrostatically actuated structures, the study of pull-in instability is necessary to determine the safe operating range, in terms of applied voltage, of an electrostatically actuated device. Collapse of its movable parts against the counter-electrode could cause malfunctioning or even damage. In fact, adhesion forces are in general strong enough to prevent positioning the device to the original configuration, moreover the electrical contact can produce high current and, due to Joule effect, local melting of the microstructure.

Sometimes pull-in phenomenon is used for the actuation itself; that is the case of deformable mirror device (DMD) for projection display developed by Texas Instruments (Senturia, 2001) and of RF switch devices (Nguyen, 2000). Measurement of pull-in voltage can be also used for an indirect experimental evaluation of Young's modulus and residual stresses in micromachined devices (Gupta, 1997).

The pull-in for deformable structure is studied in Osterberg et al.(1997), where a closed-form solution for the case of cantilever beams, double clamped beams and clamped circular diaphragms is proposed. The expression for the cantilever beam can be obtained considering the case called "bending dominated". The following expression is therefore obtained:

$$V_{\text{pull-in}} = \sqrt{\frac{0.28Et^3 g_0^3}{\varepsilon l^4 \left(1+0.42\dfrac{g_0}{w}\right)}} \tag{2-35}$$

The procedure to justify this expression is summarized in the following. It has been shown that in the case of a "single" degree of freedom system:

$$V_{\text{pull-in}} = \sqrt{\frac{8}{27}\frac{g_0^3 k}{\varepsilon A}} \tag{2-36}$$

An equivalent lumped model of the cantilever microbeam can be obtained considering a parallel plate capacitor of area $A=lw$ equal to the beam lower surface with an initial gap $g_0$ equal to the microbeam undeformed spacing and a spring element of stiffness $k_{eff}$ obtained considering the ratio between a uniform pressure load acting on the microbeam and the corresponding tip deflection:

$$k_{\text{eff}} = \frac{p_e l}{v_{\max}} = \frac{2}{3}\frac{t^3 EA}{l^4} \tag{2-37}$$

as :

$$p_e = \frac{\varepsilon w}{2} V_0^2 \left(\frac{1}{g_0^2}\right) \qquad v_{\max} = \frac{1}{8}\frac{p_e l^4}{EJ} = \frac{3}{2}\frac{p_e l^4}{Et^3 w} \tag{2-38}$$

Therefore, pull-in voltage will be:

$$V_{\text{pull-in}} = \sqrt{\frac{8}{27}\frac{g_0^3 k_{eff}}{\varepsilon A}} = \sqrt{0.20\frac{g_0^3 t^3 E}{\varepsilon l^4}} \tag{2-39}$$

Comparing the obtained expression with the expression proposed in Osterberg et al.(1997):

$$V_{\text{pull-in}} = \sqrt{\frac{0.28 g_0^3 t^3 E}{\varepsilon l^4 \left(1+0.42\frac{g_0}{w}\right)}} = 0.5292\sqrt{\frac{1}{1+0.42\frac{g_0}{w}}\frac{g_0^3 t^3 E}{\varepsilon l^4}} \tag{2-40}$$

it can be noticed that a correction term: $\sqrt{\dfrac{1.4}{\left(1+0.42\dfrac{g_0}{w}\right)}}$

is added to take into account that $k_{\text{eff}}$ evaluation was roughly approximated (upper term) and fringing field effect were not considered (lower term). This correction terms were obtained from numerical calculations. In the case of double clamped beam and of circular diaphragm a similar procedure was followed.

In case of a cantilever beam, tip displacement for which pull-in occurs can be obtained simply substituting the $V_{\text{pull-in}}$ expression in the expression of the tip displacement of the beam undergoing a uniform electrostatic pressure load:

$$v_{\text{pull-in}} = \frac{3}{4}\frac{\varepsilon l^4}{Eg_0^2 t^3} V_{\text{pull-in}}^2 = \frac{3}{4}\frac{\varepsilon l^4}{Eg_0^2 t^3}\left(0.28\frac{g_0^3 t^3 E}{\varepsilon l^4}\right)\frac{1}{\left(1+0.42\frac{g_0}{w}\right)} = 0.21 g_0 \frac{1}{\left(1+0.42\frac{g_0}{w}\right)} \tag{2-41}$$

in the plane configuration (without fringing effects): $v_{\text{pull-in}}=0.21g_0$

This result does not agree with the numerical results obtained by Cheng et al. (2004) showing, for different beam geometry a pull-in always occurring for: $v_{\text{pull-in}}=(0.45\text{-}0.47)g_0$

In Table 1 expressions of pull-in voltages and of tip displacement at which pull-in occurs are presented for the proposed methods. In particular it can be noticed that only a moderate agreement between the values of pull-in voltage exists; on the other hand the values of displacements at which pull-in occurs proposed by the different methods are highly scattered suggesting a numerical approach to find a more accurate result. Among the analytical methods only the approach proposed in Serraute et al. (1999) permits either pull-in voltage and tip displacement value at which pull-in occurs, to be evaluated. A lack of experimental results is observed in literature. From this point of view it must be pointed out that evaluation of pull-in voltage is relatively easy to be determined, on the other hand, evaluation of tip displacement at pull-in generally requires quite sophisticated optical profilometry techniques.

**Table 1.** Pull-in of a cantilever microbeam: comparison between different methods.

| | Pull-in voltage | Tip displacement at pull-in |
|---|---|---|
| Single degree of freedom | $V_{\text{pull-in}} = \sqrt{\frac{8}{27}\frac{g_0^3 k}{\varepsilon A}} = 0.5443\sqrt{\frac{g_0^3 k}{\varepsilon A}}$ | $v_{\text{pull-in}} = 0.33 g_0$ |
| Osterberg (1997) | $V_{\text{pull-in}} = \sqrt{0.28\frac{g_0^3 t^3 E}{\varepsilon l^4}} = 0.5292\sqrt{\frac{g_0^3 t^3 E}{\varepsilon l^4}}$ | $v_{\text{pull-in}} = 0.21 g_0$ |
| Taylor exp. (const.+ linear) | $V_{\text{pull-in}} = \sqrt{\frac{g_0^3 t^3 E}{\varepsilon l^4}}$ | - |
| Serraute (1999) | $V_{\text{pull-in}} = \sqrt{0.20\frac{g_0^3 t^3 E}{\varepsilon l^4}} = 0.4472\sqrt{\frac{g_0^3 t^3 E}{\varepsilon l^4}}$ | $v_{\text{pull-in}} = 0.15 g_0$ |
| Cheng(2004) | - | $v_{\text{pull-in}} = (0.45 \div 0.47) g_0$ |

# 3 Cantilever Microbeam under Electrostatic Loads: Numerical Methods

If a more accurate analysis has to be performed, a numerical approach has to be preferred. Different approaches are available in literature: methods with reduced order model, methods based on a Newton scheme, sequential field coupling approach.

## 3.1 Methods with Reduced Order Model (Lumped Parameter)

This approach was firstly suggested by Artz et al. (1992) and is based on single degree of freedom non-linear elements which are position dependent (see Fig. 4), following the relation:

$$F = \frac{c}{(g_0 - v)^2} \tag{3-1}$$

This corresponds to approximate electrostatic loads as a series of nearly parallel plate capacitor. The great advantage of this approach is that the user subroutine capability of commercially available non-linear "macro" mechanical FEM codes can be used. The method has been proposed to evaluate centre displacement of a 5mm x 5mm square diaphragm, 0.14 mm thick , with a gap of 5 μm.

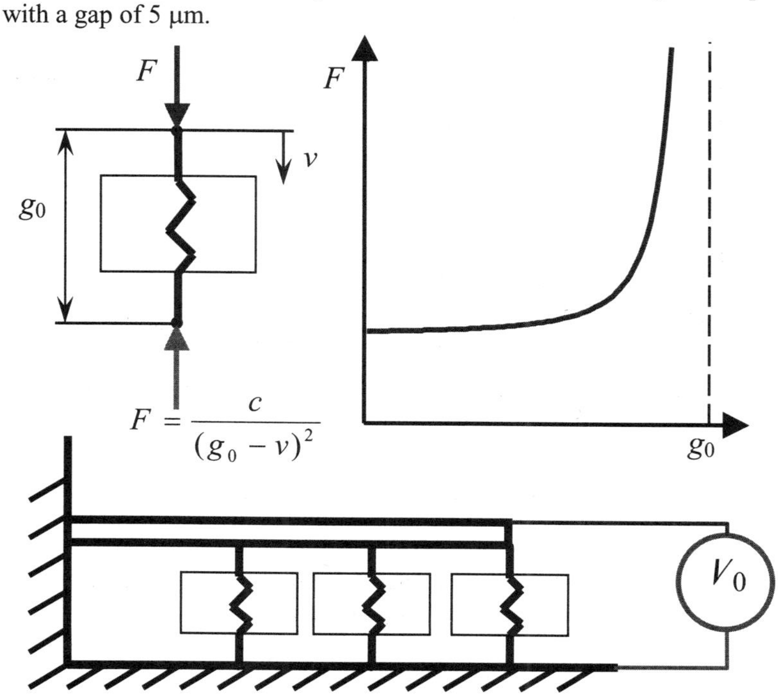


**Figure 4.** Reduced order model.

A similar approach was proposed in Gymesi (1999) where a 2 node FEM element characterized by two degrees of freedom per node, one electrical (voltage) and one mechanical (displacement) was developed. In this case the computational time is improved by referring to a so called "direct" formulation.

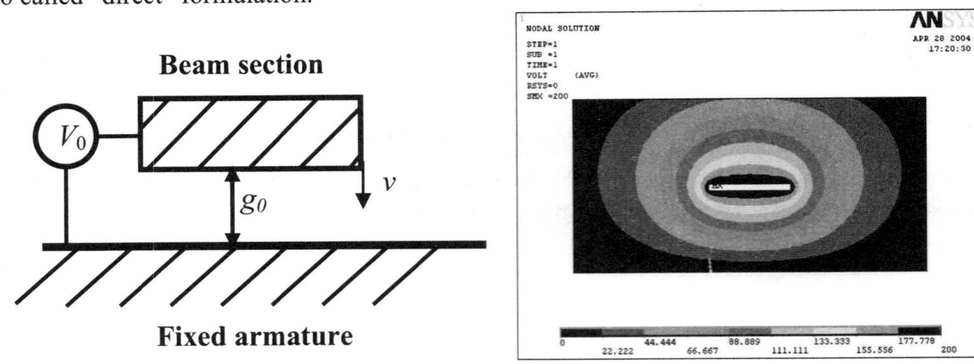


**Figure 5.** Capacitance mapping.

In practice, preliminary electrostatic analysis at different nodal displacements $v$ has to be performed, in order to achieve a discrete mapping of the capacitance $C(v)$ of the system. As shown in Figure 5, this is generally obtained considering an electrical model of the beam section at different distance from the ground.

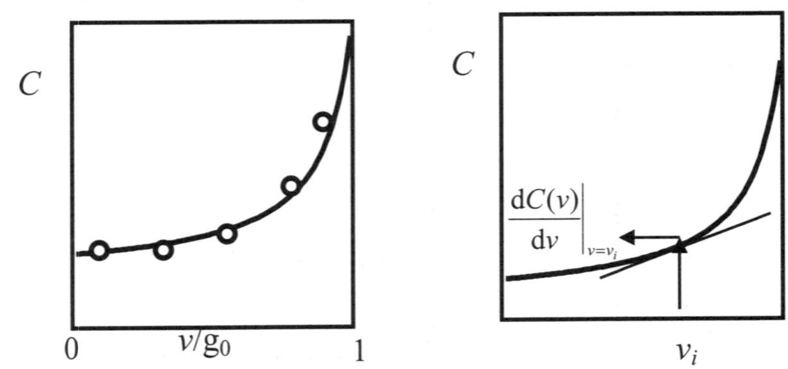


**Figure 6.** Capacitance and its derivative versus gap variation.

The curve $C(v)$ can therefore be obtained by interpolation (Figure 6), a subsequent derivation permits the electrostatic load vector to be obtained:

$$F_i = \frac{1}{2} \left.\frac{dC(v)}{dv}\right|_{v=v_i} V_0^2 \tag{3-2}$$

The relationship $F(v)$ can thus be determined. If the governing equations are considered:

$$\begin{bmatrix} [k] & [0] \\ [0] & [C(v)] \end{bmatrix} \begin{Bmatrix} \{v\} \\ \{V\} \end{Bmatrix} = \begin{Bmatrix} \{F(v)\} \\ \{Q\} \end{Bmatrix} \tag{3-3}$$

it follows that the structural analysis can be performed "directly" solving the first set of non-linear equations. The procedure is obviously profitable only if few preliminary capacitance evaluations have to be performed. This means that:

- the capacitance-displacement relationship must be the same for all the elements, i.e.

$$C_1(x) = C_2(x) = \ldots = C_n(x) \tag{3-4}$$

- the final displacement must be small, i.e. the applied voltage is small, since the capacitance-displacement relationship can be defined in few steps.

Finally it has to be enhanced that the procedure is approximated as the 3D physical domain of the electrostatic problem is modelled with 1D elements, therefore neglecting the orthogonal components of the force; on the other hand, the procedure permits fringing field effect on beam side beam to be considered.

## 3.2 Newton Scheme

As shown before, the general problem can be described by the following non-linear system of equations:

$$\begin{bmatrix} [k_{ii}] & [k_{ib}] & [0] & [0] \\ [k_{bi}] & [k_{bb}] & [0] & [0] \\ [0] & [0] & [C_{ii}(u_b)] & [C_{ib}(u_b)] \\ [0] & [0] & [C_{bi}(u_b)] & [C_{bb}(u_b)] \end{bmatrix} \begin{Bmatrix} \{u_i\} \\ \{u_b\} \\ \{V_i\} \\ V_0\{I\} \end{Bmatrix} = \begin{Bmatrix} \{0\} \\ \{F_e(V_i, V_0)\} \\ \{0\} \\ \{Q\} \end{Bmatrix} \Rightarrow$$

$$\begin{bmatrix} [k_{ii}] & [k_{ib}] & [0] & [0] \\ [k_{bi}] & [k_{bb}] & [0] & [0] \\ [0] & [0] & [C_{ii}(u_b)] & [C_{ib}(u_b)] \\ [0] & [0] & [C_{bi}(u_b)] & [C_{bb}(u_b)] \end{bmatrix} \begin{Bmatrix} \{u_i\} \\ \{u_b\} \\ \{V_i\} \\ V_0\{I\} \end{Bmatrix} - \begin{Bmatrix} \{0\} \\ \{F_e(V_i, V_0)\} \\ \{0\} \\ \{Q\} \end{Bmatrix} = 0 \Rightarrow \tag{3-5}$$

$$F(\{x\}) = 0$$

where $\{ x \}$ is the vector of all the unknowns. The system $F(x)=0$ can be solved following a Newton Raphson iterative scheme:

$$\{x_{n+1}\} = \{x_n\} + \{\Delta x_n\} \tag{3-6}$$

where $\Delta x_n$ can be obtained solving the following system of linear equations:

$$[J(x_n)]\{\Delta x_n\} = -F(x_n) \tag{3-7}$$

The approach could be tedious as the Jacobian $J(x_n)$ has to be evaluated at each iteration; to avoid this, different procedures based on conjugate gradient methods are proposed in Cai (1993) and in Gugliotta (2000). Following these approaches, the solution of the linear system:

$$[J(x_n)]\{\Delta x_n\} = -F(x_n) \tag{3-8}$$

is performed in an approximated way, as it is substituted by the non-exact condition:

$$\|[J(\{x_n\})]\{p\} + F(\{x_n\})\| \leq \eta \| F(\{x_n\})\| \qquad \text{(with: } 0<\eta<1) \tag{3-9}$$

The problem is then solved by means of multidimensional minimization methods (GCR (generalized conjugate residual) and BiCGStab (biconjugate gradient stabilized)) based on conjugate gradient methods. The general philosophy is that the minimization algorithm tries various directions $\{p\}$ until the previous equation is satisfied.
As the "direction" is imposed iteratively, explicit computation of the Jacobian $[J(\{x\})]$ is never needed, it is sufficient to approximate the product $[J(\{x\})]\{p\}$, through finite difference directional derivative:

$$[J(\{x\})]\{p\} \approx \frac{\{F(\{x\}+\alpha\{p\})\} - \{F(\{x\})\}}{\alpha} \tag{3-10}$$

It appears that each finite difference directional derivative requires $F(\{x_n\})$ to be evaluated, that is a solution of the electrical and of the mechanical problems.
Once $\{p^*\}$ vector is determined, this becomes the actual $n$-th solution according to:

$$\{\Delta x_n\} = \lambda \{p^*\} \tag{3-11}$$

The values of $\alpha$, $\eta$, $\lambda$ are determined following suitable algorithms.
In Cai (1993) the GCR scheme was adopted performing a FEM discretisation of the mechanical domain and a BEM discretisation of the electrical domain. Tests on 3D structures show that the proposed algorithm is faster and more robust then the simpler relaxation scheme. A special code CoSolve-EM (now in: Coventor Ware) was developed and commercialized.
In Gugliotta (2000) a BiCGStab scheme was adopted performing a FEM discretisation of the mechanical domain and of the electrical domain. Macros have been elaborated in the ANSYS™ parametric design language (APDL) in order to implement the algorithms.

### 3.3 Sequential Field Coupling

It is a widely used iterative approach which is based on a progressive updating of the electrostatic forces acting on the microbeam, that are corrected as the electrical field is recalculated as function of the beam deflection. The procedure is illustrated in the following, considering the numerical values presented in Figure 7.

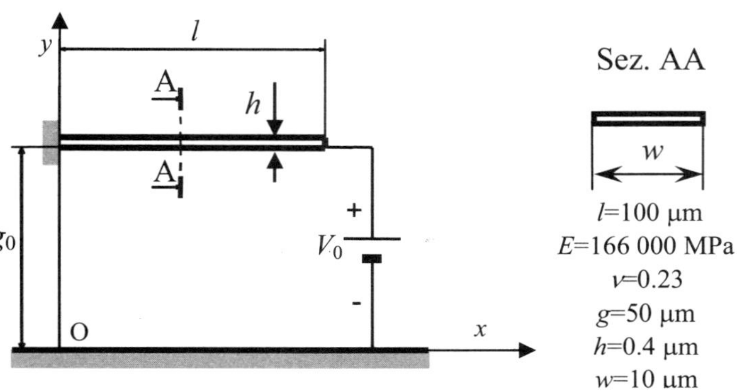


**Figure 7.** Cantilever microbeam under electrostatic loads.

As Figure 8 shows, the problem is characterized by two domains: a mechanical domain defined by the boundary 1243 and an electrical domain defined by a semi-infinite plane; the voltage is applied at the whole boundary of the microbeam. Referring to a 2D FEM modeling, the electrical domain can be described by means of triangular elements, whereas the mechanical domain can be modeled by means of beam elements (see Figure 9). It must be noticed that, in consideration of the relatively simple geometry of the mechanical part, few beam elements of equal dimensions will be sufficient for an accurate description of the mechanical behavior. On the other end, in the case of the electrical domain, a wide number of elements will be required, in particular in proximity of the beam tip where a very fine mesh will be necessary in order to correctly evaluate fringing effects.

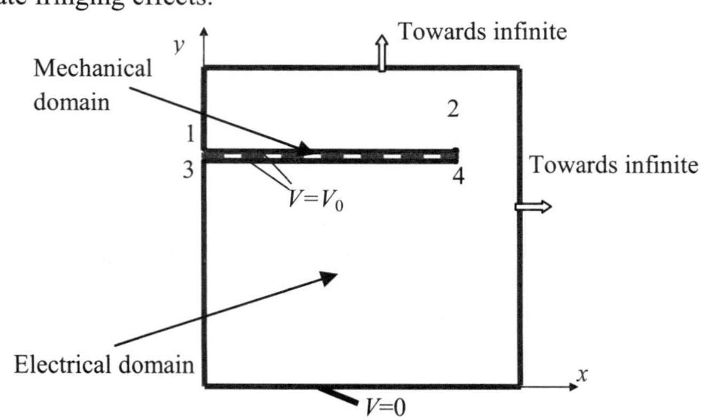


**Figure 8.** Physical domains of the problem.

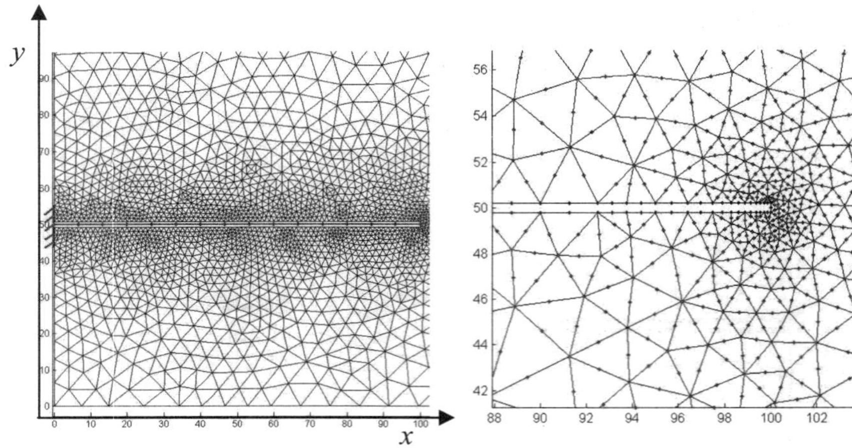

**Figure 9.** Mesh of the mechanical and of the electrical (close up view) domains.

Firstly, an electrical analysis is performed, considering the applied voltage $V_0$ and the undeformed shape of the beam. Figure 10 shows the result obtained for $V_0$=600 V.

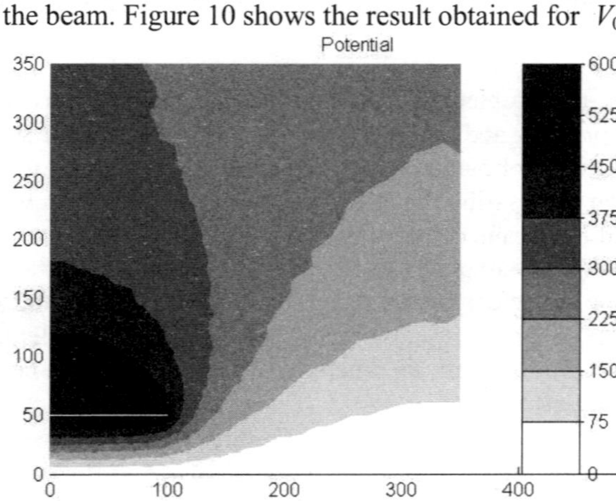


**Figure 10** Electrical solution at the first iteration: voltage distribution.

When the potential is determined in the whole domain, the vertical and horizontal components of the electrical field can be evaluated (see Fig. 11) according to the following expression:

$$\{E\}_j = \begin{Bmatrix} E_x(x,y) \\ E_y(x,y) \end{Bmatrix}_j = \begin{bmatrix} \dfrac{\partial N}{\partial x} \\ \dfrac{\partial N}{\partial y} \end{bmatrix}_j \{V\}_j = [B(x,y)]_j \{V\}_j \tag{3-12}$$

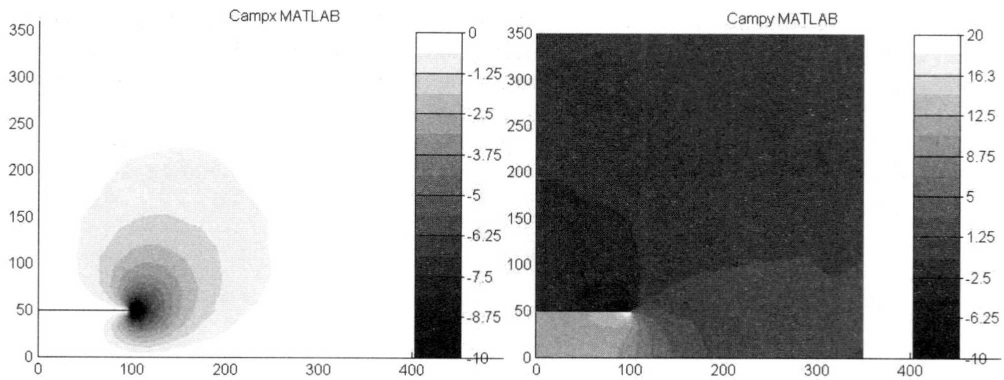


(a) (b)

**Figure 11** Electrical field components $E_x$ (a) and $E_y$ (b).

To evaluate the electrostatic loads, at first Maxwell stresses $[\sigma]$ are computed at each boundary element. For the *j*-th element:

$$[\sigma]_j = \varepsilon\left(\{E\}_j\{E\}_j^T - \frac{1}{2}\{E\}_j^T\{E\}_j[I]\right) \tag{3-13}$$

The forces acting at the interface nodes *b* can then be computed as: $\{F_{eb}\} = \iint_b [\sigma]\{n\}dS_b$ where: $\{n\}$ is the surface normal unit vector. Figure 12 shows horizontal and vertical component of the distributed electrostatic load.

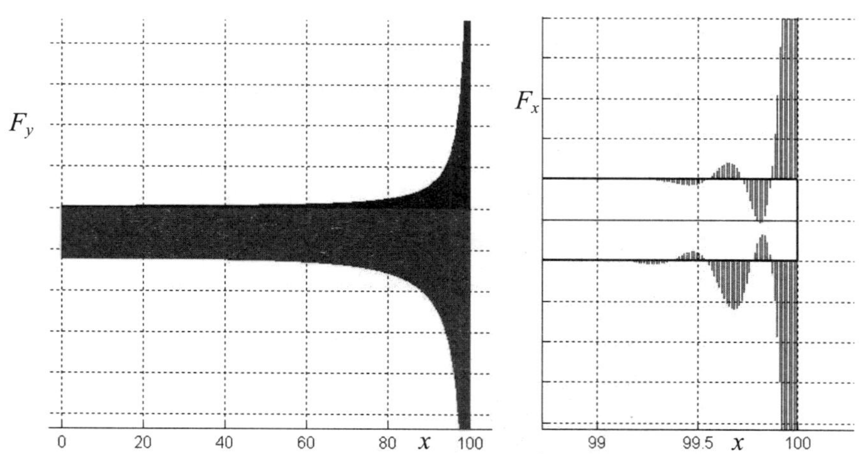


(a) (b)

**Figure 12.** Vertical (a) and horizontal electrostatic force components $F_y$ and $F_x$ and close up view (b) of distribution close to beam tip.

It can be noticed that, as expected, the vertical component is dominant, whereas the horizontal component is non null only in proximity of the tip, due to fringing effects. The vertical component resultant is 0.072 μN, corresponding to a distributed vertical load varying from 0.7 to 3 N/m. The computed forces were then applied to the nodes of the beam elements, as represented in Figure 13a; it must be enhanced that at each beam node only the resultant of the upper and lower electrostatic forces was applied. A subsequent mechanical analysis permits the beam deflection to be evaluated (Fig. 13b); in this case, at the first iteration, a tip vertical displacement of 11.45 μm was obtained.

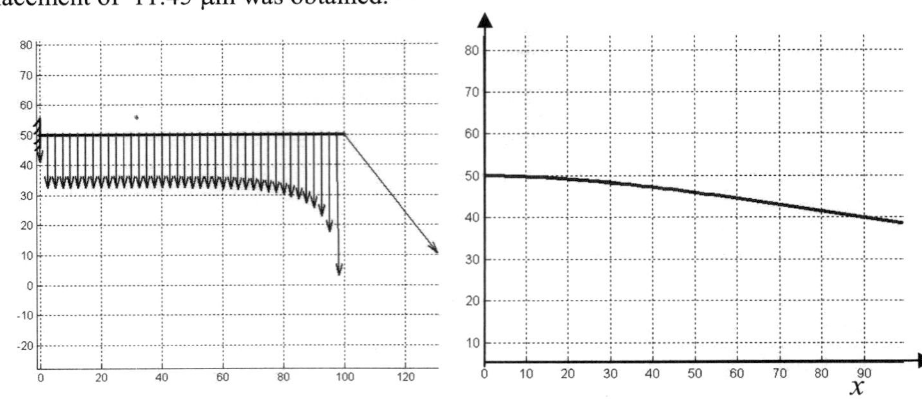


(a) (b)

**Figure 13.** First iteration: electrostatic loads acting on the beam (a) and beam deflection (b).

At this point a second iteration starts: as the electrical domain is changed, due to the deflection of the microbeam, a second mesh has to be built (see Figure 14).

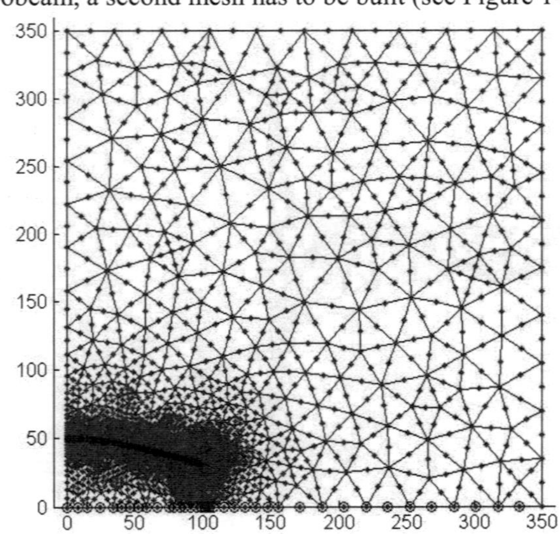

**Figure 14.** Second iteration: new electrostatic mesh.

The described procedure is repeated iteratively. From the electrical analysis, the voltage and the electrical fields components are evaluated, thus permitting the electrostatic Maxwell forces to be computed. The resultant forces are then applied at the beam nodes in the undeflected shape position: for instance in the second iteration, a value of vertical resultant of 0.0773 μN was obtained. Finally an updated beam deflection is evaluated: at the second iteration, a beam tip vertical deflection of 13.44 μm is obtained. An increment of about 15% with respect to first iteration was noticed. Further iterations are then performed unless the required tolerance is obtained, i.e. until the variation of the beam deflection is smaller than a certain value. Generally the algorithm converges quite rapidly. Obviously, the convergence depends on values of the applied voltage. Figure 15 shows the beam tip displacement versus number of iterations respectively for the case of $V$=600 V and $V$=500 V. It can be shown that in the first case the solution become stable after 10 iterations, whereas in the second case this occurs only after 6 iterations.

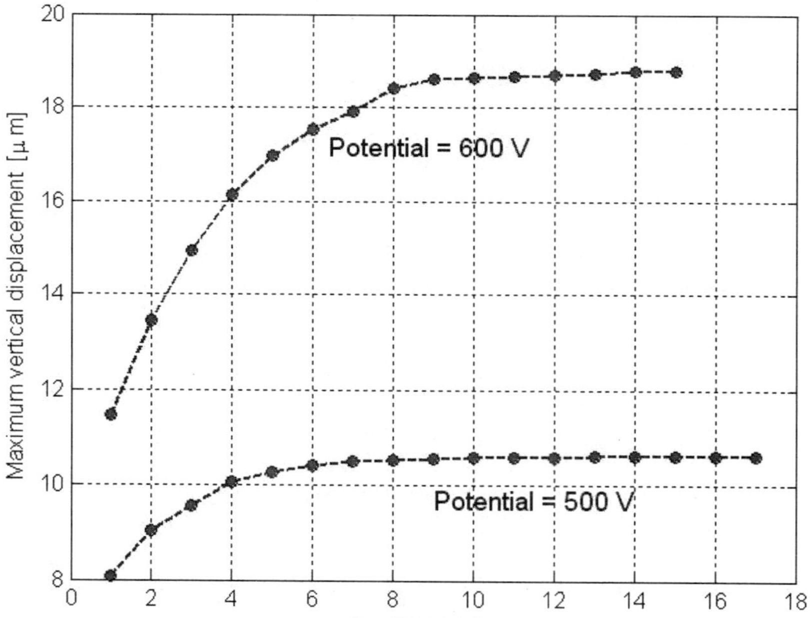


**Figure 15.** Beam tip displacement versus number of iterations.

Even if the solution is achieved in a rather limited number of iterations, the computational time could be relevant; in fact at each iteration a mechanical and electrical analysis have to be performed; moreover at each step a new mesh of the electrical domain is required. This aspect is probably the most significant from the point of view of the computational time. In order to speed up the convergence, a morphing techniques can be applied: mesh keeps its topology, that is the same number of nodes and elements, but it is deformed accordingly the deflected beam contour. Different approaches are available in literature (Gugliotta, 2001 and Gimnesy (1999).

## References

Artz, B.E., Cathey, L.W. (1992). A finite element method for determining structural displacement resulting from electrostatic forces., *Proc. IEEE Solid State Sensor and Actuator Workshop*, Hilton Head, SC, 190-193.

Bumkyoo Choi, Lovell, E. G. (1997). Improved analysis of microbeams under mechanical and electrostatic loads. *J. Micromech. Microeng.* 7:24-29.

Cai, X., Yie, H., Osterberg, P., Gilbert, J., Senturia, S., and White, J. (1993). A relaxation/multipole-accelerated scheme for shelf-consistent electromechnaical analysis of complex 3D microelectromechanical systems. *IEEE Proc. Int. Conf. on Comp. Aided Design,* (Santa Clara, CA), 283-286.

Cheng, J., Zhe, J., and Wu, X. (2004). Analytical and finite element model pull-in study of rigid and deformable electrostatic microactuators. *J. Micromech. Microeng.* 14:57-68.

Collenz, A., De Bona, F., Gugliotta, A., Somà, A., (2004). Large deflections of microbeams under electrostatic loads. *J. Micromech. Microeng.* 14:365-373.

Fukuda, T. and Menz, W., Eds. (1998). *Handbook of sensor and actuators*, Vol.6, (Amsterdam) Elsevier.

Gymesi, M., Ostergaard, D. (1999). Electromechanical transducers for MEMs analysis in Ansys. *Proc. of MSM Conf.,* (Puerto Rico), 270-273.

Gugliotta, A., Somà, A., De Bona, F., and Mola, E. (2001). Meshing approach in non-linear FEM analysis of microtructures under electrostatic loads. In: B. Curtois et al. *Design, Test, Integration and Packaging of MEMS/MOEMS 2001*, *Proc. of SPIE*, Vol. 4408, 216-225.

Gugliotta A., Soma` A., Di Mauro, S., De Bona, F. and Roccaforte, F. (2000). Non-linear analysis of beams under electrostatic loads. *Proc. of the Symposium on Design Test Integration Packaging MEMs/MOEMs*, Paris, 90-98.

Gupta, R. K. (1997). *Electrostatic pull-in test structure design for in-situ property measurement of microelectromechanical systems*, PhD dissertation, MIT, Cambridge.

Hsu Tai-Ran (2002). *MEMS & Microsystems; Design and Manufacture*, New York, Mac Graw Hill.

Madou, M. J., (2002), *Fundamentals of Microfabrication*, II Ed., Boca Raton, FL, CRC Press.

Nguyen Clark, T-C. (2000). Micromechanical circuits for communication transceivers, *IEEE BCTM 8.2*, 142-149.

Osterberg, P., Senturia, S. (1997). M-Test: a test chip for MEMs material property measurement using electrostatic actuated test structures. *J. of Microelectromechanical systems,* 6: 107-118.

Senturia, S. D. (2001). *Microsystems Design*, Boston, MA, Kluwer Academic.

Sarraute, E. and Dufour, I. (1999). Analytical modeling of beam behavior under different actuations. *J. of Modelling and Simulation of Microsystems*, 1:57-64.

# Design of Electro-Thermal Micro-Actuators: Mechanics and Electronic Position Detection

Kalin V. Lazarov [†] and Eniko T. Enikov [‡]

[†] Postdoctoral Research Associate, Advanced Microsystems Laboratory
[‡] Assistant Professor, Director of Advanced Microsystems Laboratory,
Department of Aerospace and Mechanical Engineering, University of Arizona, Tucson, AZ, USA

**Abstract** This chapter is devoted to the design of electro-thermal micro-actuators with capacitive position feedback. Analytical and finite element solutions of the electro-thermal and thermo-elastic problems are presented. A separate section is devoted to the challenging problem of displacement determination using charge sensitive amplifiers for high-precision capacitance measurements. These are illustrated with examples of electronic measurement circuits tested by the authors.

## 1 Introduction

Capacitive readout is one of the most widely used sensing modalities in MEMS applications due to its simplicity and ease of implementation. It is particularly well suited for surface-micromachined MEMS, where capacitances between moving structures are measured. Capacitive sensing is used in commercially available accelerometers (e.g., ADXL103 by Analog Devices and LIS3L02AQ by ST Microelectronics) and has been utilized to provide force (Enikov and Nelson, 2000) and position feedback for a variety of microsystems. The miniaturization of precision mechanical stages is very important in emerging applications such as ultra-compact atomic force microscopes, self-contained fiber optic aligners (Haake, 1997), and optical micro systems (Tuantranont et al., 2000). A basic structure used for in-plane translational measurements is the comb-sensor shown in Figure 1. The two sets of conductive fingers form a variable capacitor using air as a dielectric. In the configuration shown in Figure 1, the structure measures the displacement along the direction of the fingers and the finger-to-finger separation is equal. If the deflection in the perpendicular direction is of interest, a nonsymmetrical finger configuration is employed. Recently, Chu and Gianchandani (2003) demonstrated a precision integrated positioner for scanning microscopy. The device they developed utilizes V-beam thermal actuators and a capacitive feedback and was fabricated via deep reactive ion etching using Silicon-On-Insulator (SOI) wafers. A similar device fabricated using single-mask UV photolithography and an inexpensive nickel electroplating step (Enikov and Lazarov, 2003) is shown in Figure 2. This device, developed at the Advanced Microsystems Laboratory at the University of Arizona, integrates a folded-beam thermal actuator (Guckel et al., 1992; Enikov et al., 2004; Lazarov, 2004) and a movable mechanical frame coupled to a capacitive comb sensor. When current is run through the folded-beam thermal actuator, the different cross sections of the actuator arms result in different average beam

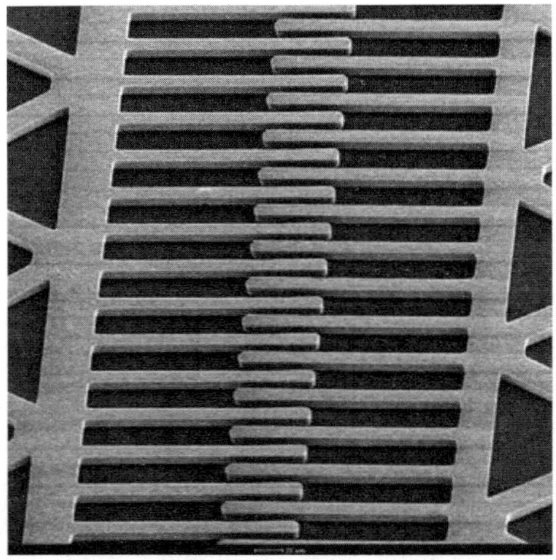

**Figure 1.** Comb structure for capacitive position sensing.

**Figure 2.** Thermal micro-actuator with integrated position sensor.

temperatures. The thin arm of the actuator is heated to a higher temperature, and the differential expansion of the beams deflects the actuator tip. The thermal actuator is very rugged and has a low CMOS compatible supply voltage, but suffers from long-term performance drift caused by the cyclic heating. The addition of the capacitive comb sensor allows closed-loop position control and eliminates the displacement uncertainty. The advantages of this device are the simplicity and low cost of the fabrication process, which has been successfully demonstrated on regular printed circuit boards (Enikov and Lazarov, 2003). Since the electrical readout design is an integral part of MEMS development, the rest of the chapter is focused on the development of simple, high-resolution modular electronics for capacitive sensing.

## 2 Mechanical and Thermal Analysis

The geometry of a common micro-actuator is shown in Figure 3. There are two types of actuators: thermal bimorphs, in which two different thermal expansion coefficient materials are used, and homogeneous actuators, in which a temperature difference is set between the narrower, "hot," and the wider, "cold," arm. In both devices, a bending moment is created in the two beams, and the two-arm structure deflects toward the beam with smaller expansion. The chosen folded-beam actuator (see Fig. 3) consists of two arms with different cross sections. When current is passed through the two arms, a temperature difference is established between the two arms. The resulting thermal strain is responsible for the lateral motion.

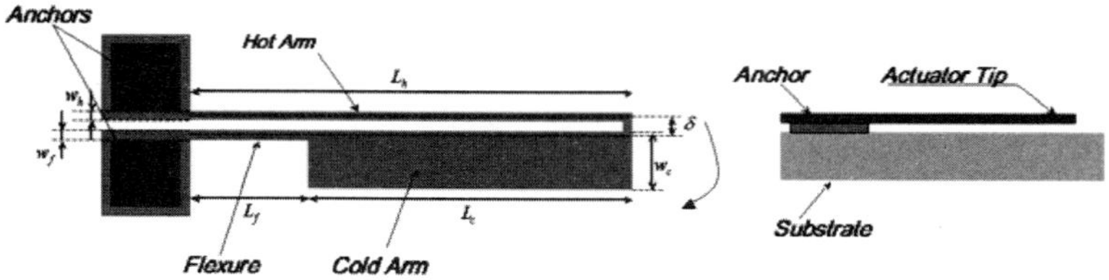


**Figure 3.** Thermal micro actuator.

The main challenge in the design of micro-actuators is to achieve sufficient actuator displacement while minimizing the power consumption. This involves electro-thermal and thermo-mechanical analysis of the folded-beam actuator. This section is devoted to this subject.

### 2.1 Thermal Analysis of Folded-beam Thermal Actuators

A thermal actuator has three modes of heat transfer: conduction, convection, and radiation. It has been estimated that radiation and convection are relatively insignificant (Hickey et al., 2003; Lott et al., 2001), i. e. almost all the heat is dissipated through conduction. Since the thermal actuator is suspended in air, the only heat dissipation path is conduction through the bonding pads, which simplified the solution of the thermal problem. A steady-state thermal analysis was conducted assuming a constant current excitation of the actuator from a current source. The resistivity of the Ni was considered a linear function of the temperature $T$, $\rho = \rho_0(1 + \beta(T - T_s))$, where the temperature coefficient of resistivity of the Ni is $\beta$, and the substrate temperature, $T_s$, was assumed constant. The differential equation describing the heat distribution along a homogeneous beam under these conditions is similar to that given by Hickey et al. (2003), however, a temperature-dependent resistivity is included

$$\frac{d^2T}{dx^2} + \rho_0(1 + \beta(T - T_s))\frac{I^2}{Kw^2t^2} = 0, \tag{1}$$

where $K$ is the thermal diffusivity of Ni and $w$ and $t$ are the width and the thickness of the arm, respectively. Q. A. Huang (2003) solved the thermal distribution problem along the unfolded actuator using similar assumptions; however, the actuators they fabricated were suspended over the substrate and the additional substrate heat flux changed the form of the solution (Enikov et al., 2004). The solution of Eq. (1) along the unfolded length of the actuator for the hot arm, cold arm, and flexure yields

$$T_h(x) = A_1 \sin(\gamma_1 x) + B_1 \cos(\gamma_1 x) + \left(T_s - \frac{1}{\beta}\right); \; x \in [0, L_h], \tag{2}$$

$$\gamma_1 = \sqrt{\frac{\rho_0 \beta}{K t^2} \frac{I}{w_h}}$$

$$T_c(x) = A_2 \sin(\gamma_2 x) + B_2 \cos(\gamma_2 x) + \left(T_s - \frac{1}{\beta}\right); \; x \in [L_h, L_h + L_c], \tag{3}$$

$$\gamma_2 = \sqrt{\frac{\rho_0 \beta}{K t^2} \frac{I}{w_c}}$$

$$T_f(x) = A_3 \sin(\gamma_3 x) + B_3 \cos(\gamma_3 x) + \left(T_s - \frac{1}{\beta}\right); \; x \in [L_h + L_c, L_h + L_c + L_f], \tag{4}$$

$$\gamma_3 = \sqrt{\frac{\rho_0 \beta}{K t^2} \frac{I}{w_f}}$$

A set of six boundary conditions is used to find the unknown coefficients $A_i$ and $B_i$. The first two conditions fix the temperature at the bonding pads equal to the substrate temperature

$$T_h(0) = T_s \tag{5}$$

$$T_f(L_h + L_c + L_f) = T_s \tag{6}$$

The last four boundary conditions enforce continuity of the temperature and the heat flux across the hot arm-cold arm and cold arm-flexure interfaces

$$T_h(L_h) = T_c(L_h) \tag{7}$$

$$w_h \frac{\partial T_h}{\partial x}(L_h) = w_c \frac{\partial T_c}{\partial x}(L_h) \tag{8}$$

$$T_c(L_h + L_c) = T_f(L_h + L_c) \tag{9}$$

$$w_c \frac{\partial T_c}{\partial x}(L_h + L_c) = w_f \frac{\partial T_f}{\partial x}(L_h + L_c) \tag{10}$$

These six boundary conditions could be used to solve for the six unknowns $A_1, B_1, A_2, B_2, A_3$, and $B_3$. Enforcing boundary condition (5) results in $B_1 = 1/\beta$. The remaining boundary conditions (6) through (10) are assembled into the linear system

$$\mathbf{A} \begin{bmatrix} A_1 \\ A_2 \\ B_2 \\ A_3 \\ B_3 \end{bmatrix} = \mathbf{b}, \tag{11}$$

where $L_1 = L_h$, $L_2 = L_h + L_c$, $L_3 = L_h + L_c + L_f$,

$$\mathbf{A} = \begin{bmatrix} 0 & 0 & 0 & \sin(\gamma_3 L_3) & \cos)\gamma_3 L_3) \\ -\sin(\gamma_1 L_1) & \sin(\gamma_2 L_1) & \cos(\gamma_2 L_1) & 0 & 0 \\ w_h \gamma_1 \cos(\gamma_1 L_1) & -w_c \gamma_2 \cos(\gamma_2 L_1) & w_c \gamma_2 \sin(\gamma_2 L_1) & 0 & 0 \\ 0 & \sin(\gamma_2 L_2) & \cos(\gamma_2 L_2) & -\sin(\gamma_3 L_2) & -\cos(\gamma_3 L_2) \\ 0 & w_c \gamma_2 \cos(\gamma_2 L_2) & -w_c \gamma_2 \sin(\gamma_2 L_2) & -w_f \gamma_3 \cos(\gamma_3 L_2) & w_f \gamma_3 \sin(\gamma_3 L_2) \end{bmatrix}, \quad (12)$$

and

$$\mathbf{b}^T = \frac{1}{\beta}[\ 1 \quad \cos(\gamma_1 L_1) \quad w_h \gamma_1 L_1 \sin(\gamma_1 L_1) \quad 0 \quad 0\ ] \quad (13)$$

Once the unknown coefficients are found, the average temperatures of the beams required for the mechanical analysis are

$$\bar{T}_i = \frac{1}{L_i} \int_{L_{i-1}}^{L_I} [A_i \sin(\gamma_i x) + B_i \cos(\gamma_i x)]\, dx + T_s - \frac{1}{\beta} \quad (14)$$

Evaluating this integral for the three beams yields

$$\bar{T}_h = \frac{1}{L_h \gamma_1} (A_1(1 - \cos(\gamma_1 L_h)) + B_1 \sin(\gamma_1 L_h)) + T_s - \frac{1}{\beta} \quad (15)$$

$$\bar{T}_c = \frac{1}{L_c \gamma_2} (A_2(\cos(\gamma_2 L_1) - \cos(\gamma_2 L_2)) + B_2(\sin(\gamma_2 L_2) - \sin(\gamma_2 L_1))) + T_s - \frac{1}{\beta} \quad (16)$$

$$\bar{T}_f = \frac{1}{L_f \gamma_3} (A_3(\cos(\gamma_3 L_2) - \cos(\gamma_3 L_3)) + B_3(\sin(\gamma_3 L_3) - \sin(\gamma_3 L_2))) + T_s - \frac{1}{\beta} \quad (17)$$

Finally, the actuator resistance, voltage drop, and power consumption are

$$\begin{aligned} R_{act} &= \frac{\rho_0(L_h + L_g}{tw_h}(1 + \beta(\bar{T}_h - T_s) + \frac{\rho_0 L_c}{tw_c}(1 + \beta(\bar{T}_c - T_s) + \frac{\rho_0 L_f}{tw_f}(1 + \beta(\bar{T}_f - T_s), \\ U_{act} &= IR_{act}, \\ P_{act} &= I^2 R_{act}, \end{aligned} \quad (18)$$

respectively. This electro-thermal analysis supplies the average beam temperatures required for the mechanical deflection analysis.

## 2.2 Finite Element Validation of the Developed Model

The analytical model developed in the previous chapter was tested against a finite element simulation obtained from ANSYS. The electro-thermal simulation was nonlinear and required coupled field analysis, with the source of nonlinearity being the change in material resistivity with temperature, according to Eq. (1). The material resistivity was specified in ANSYS in the form of a look-up table, as shown in Figure 4 (the SI resistivity value is multiplied by $10^{-6}$ because the simulations were carried out using $\mu$m as units of length). The microactuator was discretized with SOLID98 10-node tetrahedral coupled field elements with VOLT and TEMP degrees of freedom activated (KEYOPT(1)=1, the MAG degree of freedom was not used). Voltage was applied to the bonding pads as a boundary condition and the corresponding steady-state temperature distribution was obtained.

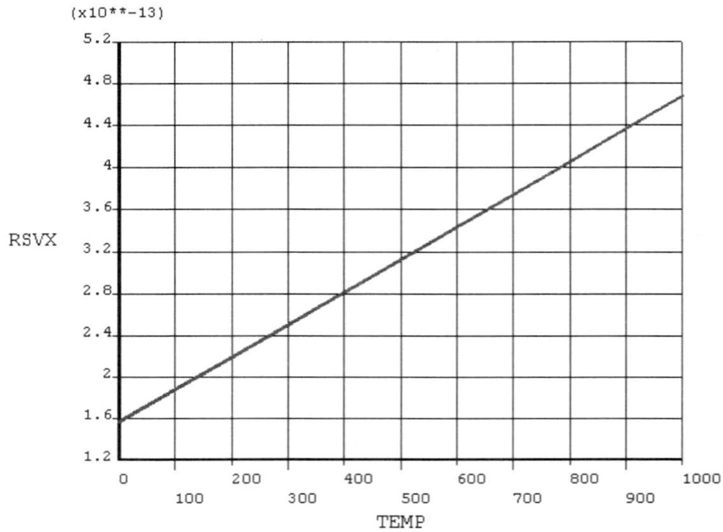


**Figure 4.** ANSYS look-up table for nickel resistivity as a function of temperature.

The ANSYS thermal model assumes that the entire heat flux is dissipated through the bonding pads, which are held at a constant temperature, equal to the substrate temperature. The current consumption is determined as a nodal reaction solution at the bonding pads. The analysis was performed with the following geometric data matching the fabricated actuators: $L_h = 1368\mu$m, $L_f = 415\mu$m, $L_c = 953$ $\mu$m, $w_h = 17.5$ $\mu$m, $w_c = 105$ $\mu$m, $w_f = 15$ $\mu$m and $\delta = 30$ $\mu$m (see Fig. 3.) The simulation was run under ANSYS 7.0, and one example of the temperature distribution for applied voltage $U = 0.2$ V is shown in Figure 5.

It can be noted that the maximum temperature is reached in the hot arm, as expected. Since the actuator dissipates heat only through the bonding pads, the cold arm stays at an elevated temperature and this offsets the spot with maximum temperature towards the actuator tip. If the actuator is suspended in close proximity above an isothermal substrate, such as a silicon wafer, the wide cold arm dissipates a significant amount of heat directly to the substrate. This reduces the cold-arm temperature and shifts the maximum-temperature spot closer to the middle of the hot arm. A comparison between the current-voltage characteristics calculated analytically from Eqs. (18) and the finite

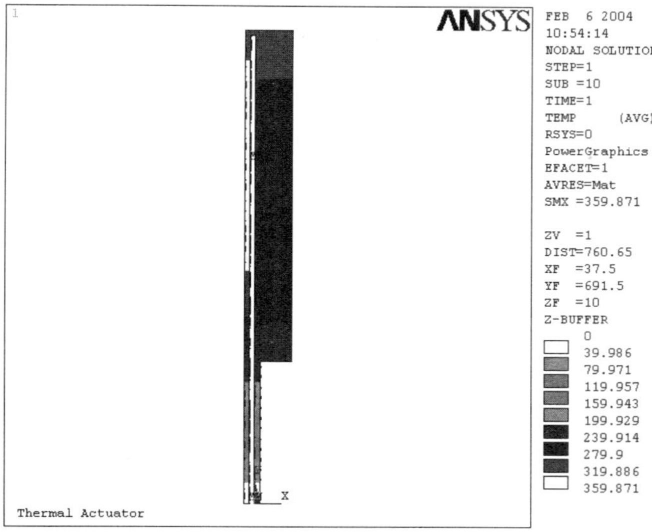


**Figure 5.** Nonlinear ANSYS solution for the temperature distribution for $U = 0.2$ V.

element simulation results is presented in Figure 6.

Two important observations could be made:

- The resistivity changes significantly due to the actuator heating – the I-V curve is far from a straight line
- The analytical model is in good agreement with the finite element simulations.

The next section describes a nonlinear mechanical model used for prediction of the free tip displacement, utilizing the average temperatures supplied by the thermal model.

## 2.3 Nonlinear Mechanical Analysis of the Folded-beam Thermal Actuator

The mechanical analysis of the folded-beam thermal actuator is essentially a two-dimensional problem. Three separate beams need to be analyzed: the hot arm, the cold arm, and the flexure. The bending of each of the beams can be described by a second-order differential equation according to Euler-Bernoulli beam theory. The resulting system of equations becomes quite complex when nonlinear geometric effects are considered. However, the equations could be simplified greatly if we note that, typically, the width of the cold arm is several times larger than the widths of the hot arm and the flexure. This results in a cold-arm cross-sectional moment of inertia two or three orders of magnitude larger, since the moment of inertia is proportional to the third power of the width. The hot and the cold arms are subjected to similar bending moments, and the larger moment of inertia of the cold arm results in its negligible bending. By assuming that the cold arm is rigid, one second-order differential equation is replaced by a linear algebraic equation, thus simplifying the analysis. The linear elongation on the other hand is proportional to the cross-sectional area of the beam, which scales linearly with

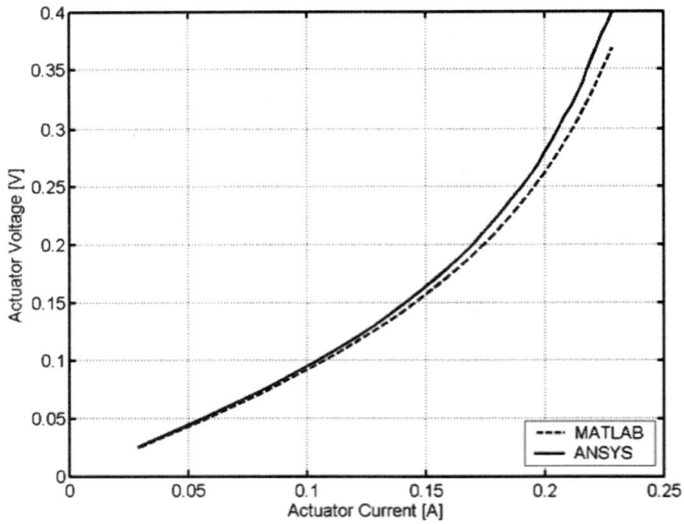


**Figure 6.** Current-voltage characteristics calculated with MATLAB and ANSYS.

the beam width. Therefore, the extension of the cold arm is the same order of magnitude and has to be accounted for. To summarize, the working assumptions for the mechanical analysis could be formulated as:

- Only the hot arm and the flexure bending is considered, provided that the cold arm is several times wider than the hot arm and the flexure
- All three beams change length due to the axial loads applied to them.

Under these assumptions, the system of equations consists of two second-order differential equations describing the bending of the hot arm and the flexure and three equations for the axial displacement of the beams. The coordinate system used for the bending of the hot arm and flexure and the deflection convention is shown in Figure 7.

The directions of the axial reaction force $P$ and the transversal force $T$ are reversed for the hot arm and the flexure (Figure 7) in order to satisfy the balance of forces. The pair of forces, $P$, separated by the distance between the arms, $\delta$, generates additional moment, and the moment balance yields

$$M_1 = P\delta - M \tag{19}$$

The general nonlinear beam equations including large displacement effects of the hot arm and flexure are

$$\begin{aligned} I_1 E w_1''(x) &= Tx - Pw_1(x) - M; \quad I_1 = tw_h^3/12, \\ I_2 E w_2''(x) &= Tx - Pw_1(x) - M; \quad I_1 = tw_f^3/12. \end{aligned} \tag{20}$$

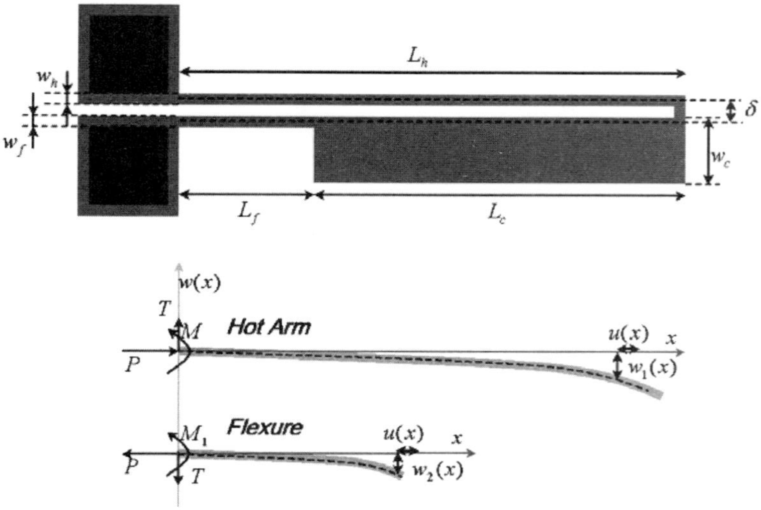


**Figure 7.** Folded-beam deflection geometry.

Here, $w_1(x)$ and $w_2(x)$ are the deflections of the hot arm and the flexure as shown in Figure 7. The built-in ends of the beams impose the following boundary conditions:

$$\begin{aligned} w_1(0) &= 0 \\ w_1'(0) &= 0 \\ w_2(0) &= 0 \\ w_2'(0) &= 0 \end{aligned} \tag{21}$$

The two second-order differential equations (20) require a total of four unknown integration constants, which could be found from the boundary conditions (21). The remaining unknowns are the reactions $T$, $P$, and $M$. The axial load $P$ causes beam deflection in the $x$-direction, shown in Figure 7 and denoted as $u$. In order to accommodate large strain effects, the mechanical strain in the beam is approximated as

$$\epsilon_x = \frac{du}{dx} + \frac{1}{2}\left(\frac{dw}{dx}\right)^2. \tag{22}$$

Using the stress defined in Eq. (22) results in the following equation for the axial force $P$

$$-P = \int_A \sigma dA = \int_A E(\epsilon_x - \alpha\Delta T)dA = \int_A E\left\{\left(\frac{du}{dx} + \frac{1}{2}\left(\frac{dw}{dx}\right)^2\right) - \alpha\Delta T\right\} dA \tag{23}$$

Integrating across the cross section of the beam yields

$$-P = EA\frac{du}{dx} + \frac{1}{2}EA\left(\frac{dw}{dx}\right)^2 - \alpha\Delta TEA. \tag{24}$$

Here, $A$ is the cross-sectional area of the beam, $E$ is the Young's modulus, and $\Delta T$ is the difference between the average beam temperature and the substrate temperature $T_s$. Integrating once again along the length of the beam results in

$$-\frac{PL}{EA} - \Delta u - \alpha \Delta T L + \frac{1}{2}\int_0^L \left(\frac{dw}{dx}\right)^2 dx. \tag{25}$$

Substituting the values for the hot arm, cold arm, and flexure using $P$ with the appropriate sign (positive for the hot arm and negative for the flexure and the cold arm) defines a system of three nonlinear equations

$$-\frac{PL_h}{EA_h} - \Delta u_h - \alpha(\bar{T}_h - T_s)L_h + \frac{1}{2}\int_0^{L_h} \left(\frac{dw_1}{dx}\right)^2 dx, \tag{26}$$

$$-\frac{PL_f}{EA_f} - \Delta u_f - \alpha(\bar{T}_f - T_s)L_f + \frac{1}{2}\int_0^{L_f} \left(\frac{dw_2}{dx}\right)^2 dx, \tag{27}$$

$$-\frac{PL_c}{EA_c} - \Delta u_c - \alpha(\bar{T}_c - T_s)L_c + \frac{1}{2}\int_0^{L_c} \left(\frac{dw_c}{dx}\right)^2 dx. \tag{28}$$

The integral of the square of the derivative of the deflection with respect to $x$ in Eqs. (26) and 27) is a known function of the reactions $T$, $P$, and $M$, according to (20 and 21). The similar term in the third equation (28) is also easy to calculate since it can be assumed that the cold arm is rigid and the cold arm rotation angle $\theta = dw_c/dx$ is constant, known function of the reactions (not a new unknown). Equation (28) can therefore be rewritten in the form

$$\frac{PL_c}{EA_c} = \Delta u_c - \alpha(\bar{T}_c - T_s)L_c + \frac{1}{2}\theta^2 L_c. \tag{29}$$

The three newly introduced equations (26)-(28) include the additional unknowns $\Delta u_h$, $\Delta e_f$, and $\Delta u_c$. In order to complete the system of equations, it is necessary to define three additional conditions to solve for the remaining unknown reactions. The first such condition is the previously mentioned cold-arm rigidity condition. The assumption that the cold arm does not bend forces the slope at the end of the hot arm and the slope at the end of the flexure to be equal, which can be expressed as

$$w_1'(L_h) = w_2'(x) = \theta. \tag{30}$$

Since the hot arm is connected to the cold arm and the flexure, the transversal tip displacement $w$ calculated from the hot arm must be equal to the deflection calculated from the flexure and the cold arm. The second boundary condition is therefore

$$w_1(L_h) = w_2(L_f) + w_1'(L_h)L_c. \tag{31}$$

The last boundary condition states that the extension of the hot arm must be equal to the extension of the cold arm and the flexure

$$\Delta u_h - \Delta u_g = \Delta u_f + \Delta u_c. \tag{32}$$

The additional term $\Delta u_g$ is the geometric displacement caused by the rotation of the small beam with length $\delta$ connecting the hot arm to the cold arm. The geometric displacement translates the displacement of the hot arm along its axis to the axis of the flexure and the cold arm and is

$$\Delta u_g = -\delta \sin(\theta) \approx -\delta\theta \approx -\delta w_2'(L_f) \qquad (33)$$

Substituting Eqs. (26), (27), (29) and (33) in (32) yields

$$\begin{aligned} \frac{P}{E}\left(\frac{L_f}{A_f}+\frac{L_c}{A_c}+\frac{L_h}{A_h}\right) &= \alpha(L_h\Delta T_h - L_f\Delta T_f - L_c\Delta T_c) \\ &-\frac{1}{2}\int_0^{L_h} w_1'^2 dx + \frac{1}{2}\int_0^{L_f} w_2'^2 dx + \frac{1}{2}L_c\theta^2 + \delta\theta. \qquad (34) \end{aligned}$$

The last condition (34) completes the system of equations. The final set of equations consists of the two second-order differential equations for the deflection (20), the four boundary conditions enforced by the built-in ends of the beams (21), and the three boundary conditions (30), (31), and (34). The detailed solution and analysis of these had been provided by Lazarov (2004). Figure 8 shows a comparison between the analytical and finite element models described above and the experimental measurements made on the same actuator. As expected, the error increases at higher current levels due to the nonlinear temperature dependence of the resistivity.

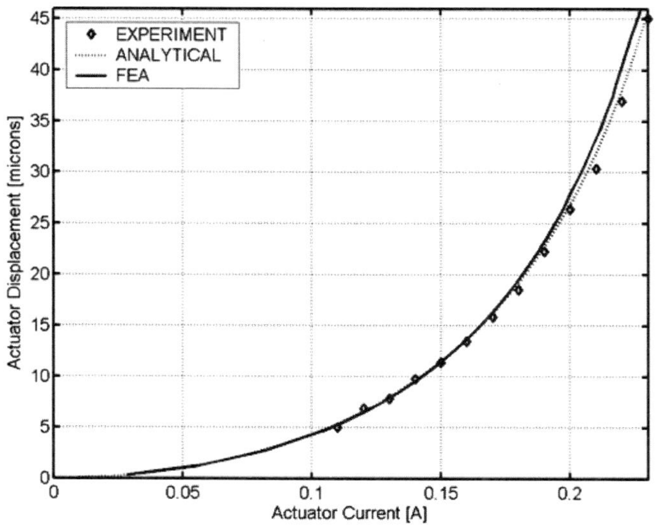


**Figure 8.** Displacement vs. current: simulation and experiment.

Contrary to the commonly analyzed case of sacrificially released actuators, if the actuators are suspended in air, the cold arm remains at an elevated temperature. The significance of the wider cold arm is therefore two-fold: (i) to decrease the electrical resistance and generate less heat and (ii) to decrease the thermal resistance by increasing the conduction cross section. Ideally, the cold arm will be at a temperature close to the substrate temperature, thus maximizing the thermal difference between the arms. If the entire heat flux is conducted to the substrate, increasing the cold arm's width above $w_c \approx 5w_h$ is not beneficial since at this point the thermal "bottleneck" is the cold arm cooling through the flexure. By fixing the cold arm's width to $w_c \approx 5w_h$, the design optimization requires a balance between two conflicting thermal and mechanical requirements:

- Long flexure provides a mechanically softer structure and produces larger free tip displacement for a given thermal strain. However, the added thermal resistance of the flexure keeps the cold arm at a high temperature and decreases the temperature difference
- Short flexure enhances the cold arm cooling and increases the temperature difference, but makes the actuator mechanically stiffer.

## 3 Capacitive Readout of Position

Accurate positioning of a MEMS device requires closed-loop control. Most commonly, the position in such cases is measured via capacitance measurements. The major difficulty in the measurement of MEMS capacitors is the presence of parasitic capacitances associated with the signal traces connecting the MEMS capacitive sensor. A schematic of the equivalent circuit showing the parasitic capacitances $C_1$ and $C_2$ and the mechanical feedback capacitance $C_{\mathrm{MF}}$ is shown in Figure 9.

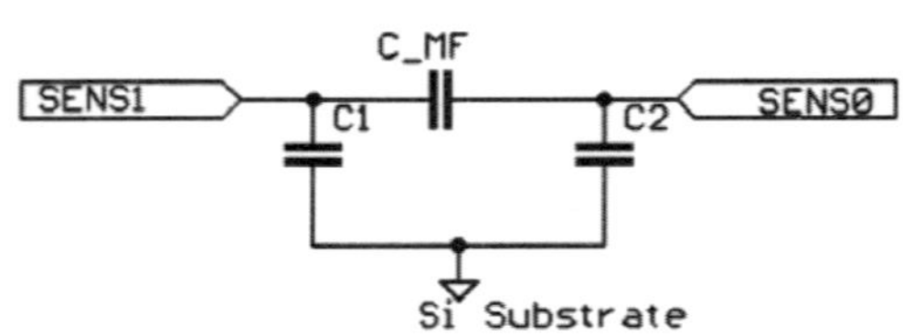


**Figure 9.** Equivalent circuit of the capacitive position sensor

The figure depicts a common situation, where the MEMS structures are built on an oxidized low-resistivity silicon substrate. Connecting the substrate to the common electrical ground avoids crosstalk between different sensors. Typically, the signal traces leading to the MEMS capacitive sensor (comb structure or parallel plate capacitor have significant capacitance, since the insulating silicon oxide is thin, on the order of 1 $\mu$m, and $\epsilon_{\mathrm{SiO_2}} = 4.2$). For example, the fabricated device shown in Figure 9 is anchored to the substrate via relatively large bonding pads, 500 $\mu m \times 1000$ $\mu m$, and $C_1$ and $C_2$ are approximately 1900 $fF$. In contrast, the capacitive comb structure, shown in the same figure, has 45 fingers, a 50-$\mu$m finger overlap, a 4-$\mu$m air separation, and is 14-$\mu$m high, which yields a mechanical feedback capacitance of $C_{\mathrm{MF}} = 140\ fF$. Some applications require even longer signal traces and bigger bonding pads – an example is a capacitive MEMS structure built on a flexible substrate, with the rigid readout electronics located at the end of the flex-circuit. It is clear that the parasitic capacitances are quite large compared to the sensing capacitor and that they need to be removed from the signal acquisition path.

An elegant way of eliminating the parasitic capacitances to ground is to measure the charge flowing through the sensor capacitor during voltage transitions. This method is essentially a two-port force-sense measurement (Smartec, 2003), where a square-wave voltage excitation is applied on the left-hand side of the circuit in Figure 9 and the charge flowing through $C_{\mathrm{MF}}$ is integrated with a charge-sensitive amplifier.

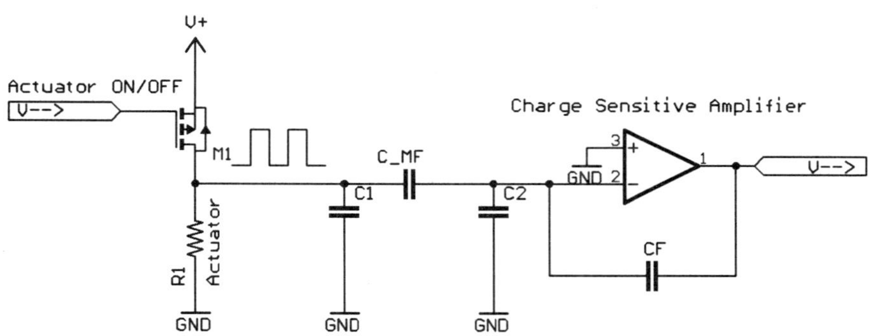


**Figure 10.** Elimination of the parasitic capacitances using a two-port measurement with voltage excitation and charge sensitive amplifier

This concept is illustrated in Figure 10, where the square-wave excitation comes from the voltage drop across the thermal actuator itself. In this case, the charge-sensitive amplifier is realized with an operational amplifier and a feedback capacitor, $C_F$. This technique works if the power to the thermal actuator is supplied by a pulse-width modulation, which has the advantage of very high electrical efficiency. The described charge measurement method could also be used in the more general case where a separate lead is available for the capacitive sensor, and in this case an independent voltage driver, not connected to the actuator, supplies the voltage waveform. The low output impedance of the waveform generator (or the low resistance of the thermal actuator in Figure 10) is connected in parallel with $C_1$ and eliminates this parasitic capacitance. The high open-loop gain of the amplifier, on the other hand, keeps the negative input of the amplifier connected to $C_2$ at constant potential, thus preventing this capacitor from charging and eliminating it from the signal acquisition chain. Ideally, if the open-loop gain of the amplifier is infinite, all the charge injected through the sensor capacitance $C_{\mathrm{MF}}$ is integrated in the feedback capacitor of the charge-sensitive amplifier. For real circuits, a charge collection gain, $G_Q$ , is defined as (Spieler, 2001)

$$G_Q = \frac{Q_F}{Q_F + Q_{\mathrm{in}}} \tag{35}$$

Here, $Q_{\mathrm{F}} = V_{\mathrm{out}}C_{\mathrm{F}}$ is the charge captured by the charge-sensitive amplifier and $Q_{\mathrm{in}} = V_{\mathrm{in}}(C_{\mathrm{in}}+C_2)$ is the charge lost in the parasitic capacitance $C_2$ and the input capacitance of the amplifier $C_{\mathrm{in}}$. The open-loop gain of the amplifier, $A_{\mathrm{OL}}$ , defined as $A_{\mathrm{OL}} = V_{\mathrm{out}}/V_{\mathrm{in}}$, transforms Eq. (35) to

$$G_Q = \frac{V_{\mathrm{out}}C_F}{V_{\mathrm{out}}C_F + \frac{V_{\mathrm{out}}}{A_{\mathrm{OL}}}(C_{in} + C_2)} = \frac{1}{1 + \frac{(C_{in}+C_2)}{A_{\mathrm{OL}}C_F}} \tag{36}$$

The benefit of the high-gain amplifier in Figure 10 is two-fold: (i) nearly all the charge generated from the voltage transition across the measured capacitor is measured, i.e., $G_Q \approx 1$ ,and (ii) the gain $G_Q$ does not depend on the parasitics. The second property is quite important, since the parasitic capacitances can vary widely due to variations in the fabrication process while the sensor capacitance is more accurate. Assuming that $G_Q \approx 1$, the charge generated during a voltage transition with amplitude $V_T$ across the sensing capacitance $C_{\mathrm{MF}}$ is $Q_{\mathrm{inj}} = V_T C_{\mathrm{MF}}$ and the output of the charge sensitive amplifier is

$$V_{\mathrm{out}} = -V_T \frac{C_{MF}}{C_F} \tag{37}$$

The negative sign in Eq. (37) is due to the fact that the inverting input of the operational amplifier in Figure 10 is connected to the capacitive sensor.

The described charge integration method is used in some of the most sensitive commercial integrated circuits for MEMS capacitive measurements, such as UTI (Smartec Inc.) and MS3110 (Micro Sensors Inc.). Even though these circuits are high resolution and easy to use, they have limited bandwidth and cannot handle the low resistance of the thermal actuator connected to ground (Figure 10).

One possible solution is to append a pulse-shaping amplifier and a peak detector to the circuit, as shown in Figure 11. The pulse-shaper acts like a differentiator and outputs a pulse during the rising and falling edges of the square wave coming from the CSA. The difference between the maximum (or minimum) amplitude of one of these pulses and the baseline voltage (the output voltage when there are no pulses) is linearly proportional to the charge injected in the CSA. Since a single voltage pulse across the measured capacitor results in a square wave output of the CSA, and the pulse shaper generates one positive and one negative pulse, the total difference between the minimum and the maximum voltages of the pulse is directly proportional to the measured MEMS capacitance. Finally, the peak detector is used to hold the maximum and the minimum values of the voltage pulse.

The main advantage of this approach compared to off-the-shelf components, such as UTI and MS3110, is the significant increase in the measurement speed, without sacrificing accuracy. A simple, low-cost system consisting of a few operational amplifiers implementing the circuit in Figure 11 can take a single measurement in less than 5 $\mu s$ with input charge noise of about 600 electrons. Furthermore, the design is very flexible, and the overall system gain can be adjusted in a very wide range (three orders of magnitude or more) by simply adding more amplifier/filter stages.The charge-sensitive amplifier, pulse-shaper, and peak detector are described in detail in the following.

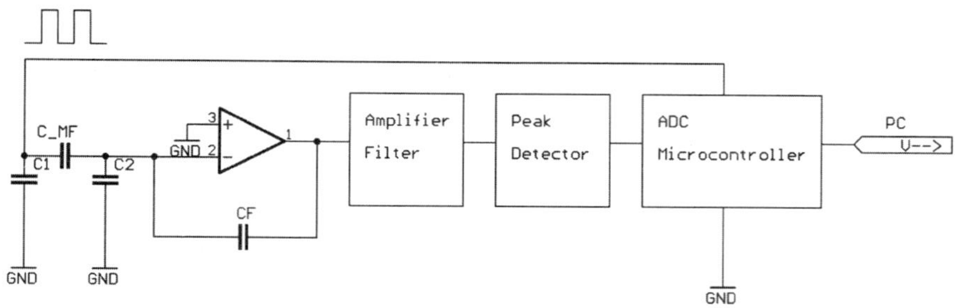


**Figure 11.** Block diagram of the MEMS capacitor measurement system

# 4 Electrical Circuit Design

## 4.1 Charge-Sensitive Amplifier

A practical implementation of a charge sensitive-amplifier (CSA) using an operational amplifier is shown in Figure 12. The operational amplifier is powered by a single +5 $V$ power supply and the resistor divider, $R_1$ and $R_2$, filtered by $C_1$ biases the positive input to half the supply voltage. To reduce the noise, the power supply of the operational amplifier is filtered with 10 $\mu F$ tantalum and 100 $nF$ ceramic capacitors, $C_2$ and $C_3$. The

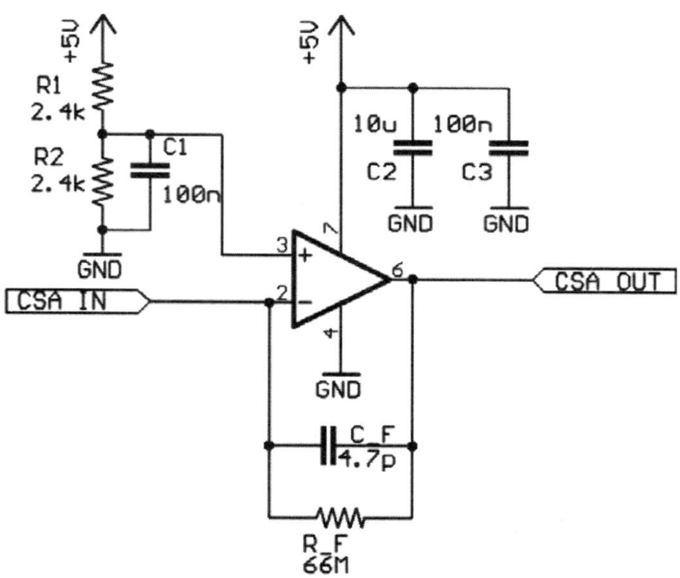


**Figure 12.** Practical realization of CSA

feedback capacitor $C_F$ is a high-accuracy temperature-compensated ceramic type and is connected in parallel with a high-value feedback resistor $R_F = 66\ M\Omega$. The feedback resistor provides a path for the input bias current of the opamp and acts as a reset circuit by discharging the feedback capacitor. The operational amplifier used in Figure 12 must have a very low input bias current ($\approx pA$), and CMOS and JFET input amplifiers are a must. Furthermore, the opamp must have low input voltage noise and a high slew rate. Some examples of devices satisfying these requirements are MAX4477 from Maxim Integrated Products and OPA134 from Texas Instruments. Monolithic charge-sensitive amplifiers are also available off-the-shelf, such as CR110 from Cremat Inc., and H4083 from Hamamatsu.

### 4.2 Pulse-Shaping Amplifier

The pulse-shaping amplifier filters and amplifies the signal coming from the CSA; one possible realization is shown in Figure 13. For simplicity, the filter is shown in a configuration for a split power supply. The single supply operation requires biasing of the non-inverting amplifier input, as shown in Figure 12.

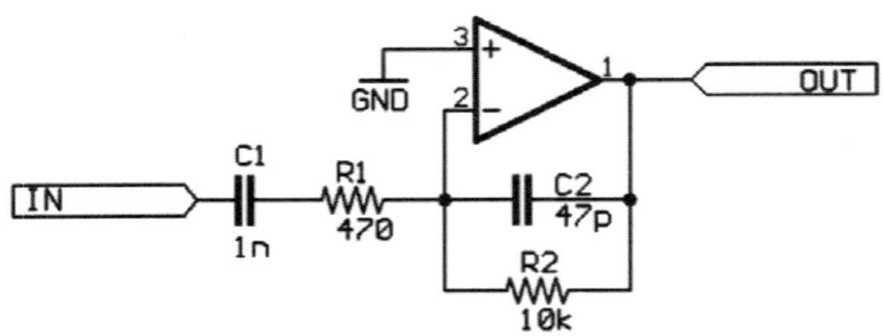


**Figure 13.** Amplifier/shaper stage.

Assuming that the operational amplifier has infinite open-loop gain and input impedance, the input current $i_{\text{in}}$ flowing through $R_1$ and $C_1$ creates a voltage drop across $R_2$ and $C_2$, which is the output voltage of the filter. The filter transfer function can be derived by equating the input and output currents

$$i_{\text{in}} = \frac{V_{\text{in}}}{R_1 + \frac{1}{jwC_1}} = -V_{\text{out}}\left(\frac{1}{R_2} + jwC_2\right) \tag{38}$$

$$\frac{V_{\text{out}}}{V_{in}} = -\frac{jwC_1R_2}{(1 + jwC_1R_1)(1 + jwC_2R_2)} \tag{39}$$

The filter transfer function is therefore

$$H(s) = -\frac{sC_1R_2}{(1 + sC_1R_1)(1 + sC_2R_2)} \tag{40}$$

The transfer function of the stage has one zero and two poles and amplifies the input signal. If the gain of a single stage is not sufficient, multiple stages can be cascaded. However, since the desired overall transfer function has a Gaussian profile with a transfer function close to $H(s) = Ks/((s+a)^n)$, $n \approx$ 3-4, the additional stages have to contribute poles only. This could be accomplished by moving one of the poles in Eq. (40) to low frequencies, effectively canceling the zero in the numerator. The pole can be moved by increasing the value of the decoupling capacitor $C_1$ in the stages after the first one, and values around $C_1 = 1\ \mu F$ were found to work well.

### 4.3 Peak Detector

The last stage in the analog signal conditioning is the peak detector circuit shown in Figure 14. The positive peak detector consists of $Q_1$, $Q_2$, $R_1$, $R_2$, and $C_1$. The first transistor $Q_1$ and its load resistor $R_1$ buffer the input voltage for the second peak-detecting stage, realized with $Q_2$, $R_2$, and $C_1$. The hold capacitor $C_1$ is charged to the

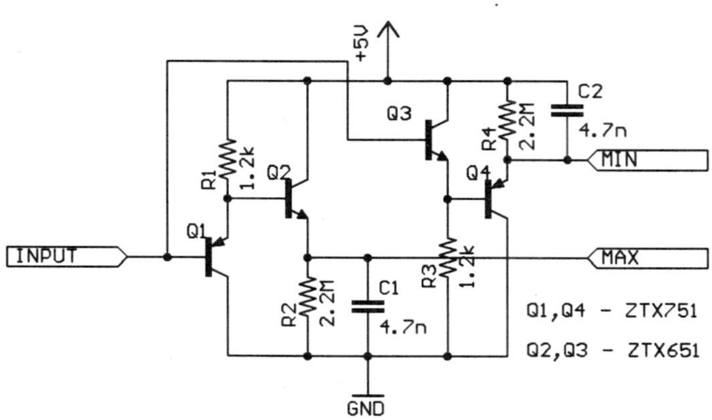


**Figure 14.** Positive/negative peak detector.

maximum of the input voltage through the rectifying base-emitter junction of $Q_2$, which is equivalent to the classic peak detector with a diode and a capacitor. The emitter resistor $R_2$ sets the DC bias current of $Q_2$. The operation of the negative peak detector consisting of $Q_3$, $Q_4$, $R_3$, $R_4$, and $C_2$ is completely analogous. The hold capacitors are discharged and the system reset to the original position after each pulse by connecting $C_1$ and $C_2$ to ground and the supply voltage, respectively, through lower-value resistors ($\approx 1k\Omega$).

### 4.4 Analog-to-Digital Converter and Microcontroller

The last part of the block diagram in Figure 11 is the microcontroller, with an integrated analog-to-digital converter (ADC). The microcontroller generates the square-wave testing waveform, discretizes the output of the peak detectors with its internal ADC, and resets the hold capacitors. Finally, it calculates the value of the sensor capacitance from the measured voltages and transmits the information to a PC or PDA, which completes the measurement system.

## Acknowledgement

This material is based upon the work supported in part by the National Science Foundation under Grant No. EEC-0407369. Any opinions, findings, and conclusions or recommendations expressed in this material are those of the author(s) and do not necessarily reflect the views of the National Science Foundation.

## Bibliography

Chu, L.L. and Gianchandani, Y.B. (2003) A micromachined 2d positioner with electrothermal actuation and sub-nanometer capacitive sensing. *J. of Micromechanics and Microengineering*, 13, pp. 279–285.

Enikov, E.T. and Lazarov, K.V. (2003) Pcb-integrated metallic thermal micro-actuators. *Sensors and Actuators A*, 105, pp. 76–82.

Enikov, E.T., Lazarov, K.V., and Kedar, S.S. (2004) Analytical and experimental analysis of folded beam and v-shaped thermal microactuators. In *SEM 10th International Congress and Exposition*. Costa Mesa, California USA.

Enikov, E.T. and Nelson, B. (2000) Three dimensional microfabrication for multi-degree of freedom capacitive force sensor using fiber chip coupling. *J. of Micromechanics and Microengineering*, 10, pp. 492–497.

Guckel, H., Klein, J., Christenson, T., Skrobis, K., Laudon, M., and Lovell, E.G. (1992) Thermo-magnetic metal flexure actuators. In *Solid-State Sensor and Actuator Workshop*. Hilton Head.

Haake (1997) *Fiber optic connector having at least one microactuator for precisely aligning an optical fiber and an associated fabrication method.* US Patent 5,606,635.

Hickey, R., Sameoto, D., Hubbard, T., and Kujath, M. (2003) Time and frequency response of two-arm micromachined thermal actuators. *J. of Micromechanics and Microengineering*, 13, pp. 40–46.

Lazarov, K.V. (2004) *Design, Fabrication and Control of Hybrid Thermal/Piezoelectric MEMS Array*. Ph.D Thesis, The University of Arizona.

Lott, C.D., McLain, T.W., Harb, J.N., and Howell, L.L. (2001) Time and frequency response of two-arm micromachined thermal actuators. In *Technical Proceedings of the 2001 International Conference on Modeling and Simulation of Microsystems.*

Q. A. Huang, N.K.S.L. (2003) Analysis and design of polysilicon thermal flexure actuator. *J. of Micromechanics and Microengineering*, 9, pp. 64–70.

Smartec (2003) *Startup guide for measurement of capacitors with three signal technique.*

Spieler, H. (2001) *Lecture notes on Radiation Detectors and Signal Processing - II. Signal Formation, University of Heidelberg.*

Tuantranont, A., Bright, V.M., Zhang, J., Zhang, W., Neff, J., and Lee, Y.C. (2000) Mems-controllable microlens array for beam steering and precision alignment in optical interconnect systems. In *Solid-State Sensor and Actuator Workshop*. Hilton Head, pp. 101–104.

# Design of Compliant Micromechanisms

Francesco De Bona[1] and Saša Zelenika[2 *]

[1] Dipartimento di Ingegneria Elettrica, Gestionale e Meccanica, University of Udine, Udine, Italy
[2] University of Rijeka – Faculty of Engineering, Rijeka, Croatia

**Abstract.** A broad overview of the topics related to the mechanical design of compliant micromechanisms is presented. Design methodologies to be used in the design of devices based on leaf springs, flexural notches and continuum structures with distributed compliance are given, and a critical presentation of the peculiarities of these solutions is provided. The extensive bibliographical list is given as means to extend further the study to details of each of the treated topics.

# 1 Introduction

A compliant mechanism is a device that gains at least part of its mobility from the deflection of flexible members rather than from moveable joints only (Howell, 2001) – Fig. 1.

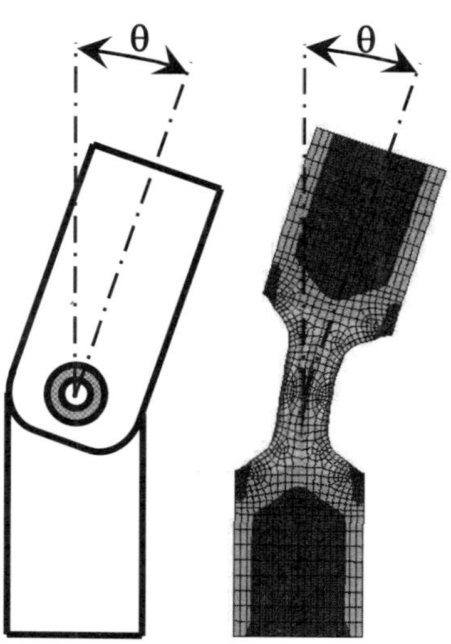


**Figure 1.** Rotation achieved by using a conventional mechanical and a compliant joint.

* The authors wish to thank M. Gh. Munteanu of the University of Brasov, Romania, for his contribution in some of the illustrated work.

Compliant mechanisms are widely used in the field of precision engineering as well as micro and nanotechnologies (see e.g. Howell, 2001, Jones, 1988, Lobontiu, 2003, Lobontiu and Garcia, 2005, Madou, 2002, Senturia, 2001, Slocum, 1992, Smith and Chetwynd, 1992, Smith, 2000, and Trylinsky, 1971).

In fact, if compared with conventional mechanical bearings, flexible bearings allow the following advantages to be achieved:

- high precision
- absence of friction
- limited hysteresis
- absence of wear
- no need for lubrication
- no risk of jamming
- absence of backlash
- reduction in weight
- suitable for operation in harsh or special environments (including clean-room microproduction facilities)
- main sources of errors are systematic and therefore simple control laws can be used
- reduced costs
- possibility of monolithic manufacturing and thus of "design for no-assembly" and of parallel instead of serial kinematics.

In particular in the case of microsystems where the microfabrication processes are based on lithographic techniques, solutions based on monolithic configurations can be easily adopted, thus overcoming the expensive phase of packaging and microassembly (Yang and Nelson, 2004).

On the other hand solutions based on compliant mechanisms evidence the following drawbacks:

- limited strokes
- limited load capacity
- presence of restoring forces
- complex kinematics
- presence of parasitic motions, i.e. displacements along the secondary degrees of freedom (see below).

The effects of such limitations can be minimised if an accurate model of the compliant devices comprising mechanical non-linearities is adopted.

From the mechanical design point of view, compliant devices can be classified as:

- compliant mechanisms based on leaf springs,
- compliant mechanisms employing flexural hinges (localised compliance) and
- continuum structures with distributed compliance.

In the following paragraphs the main aspects of the design of each of these solutions will be outlined.

## 2 Compliant Micromechanisms Based on Leaf Springs

Mechanisms based on leaf springs are devices in which the motion is obtained by coupling the movable member with the rigid one by means of beam-shaped flexible elements characterized by a marked bending compliance and a high stiffness along the other degrees of freedom.
Such devices follow different configurations:

1. Single leaf spring devices: a cantilever or a double clamped beam used in off-the-shelf ultra-high precision measurement devices such as the atomic force microscope (AFM), the scanning tunnelling microscope (STM), the radio frequency micro-electromechanical systems (RF-MEMS) and the electrostatic projection displays, as well as in several custom developed nanotechnological solutions such as those for biological applications (micro balance) – Bhushan (2004).
2. Two-beam positioning devices (Fig. 2) used to achieve either a translation (parallel spring translators) or a rotation (cross-spring pivots).

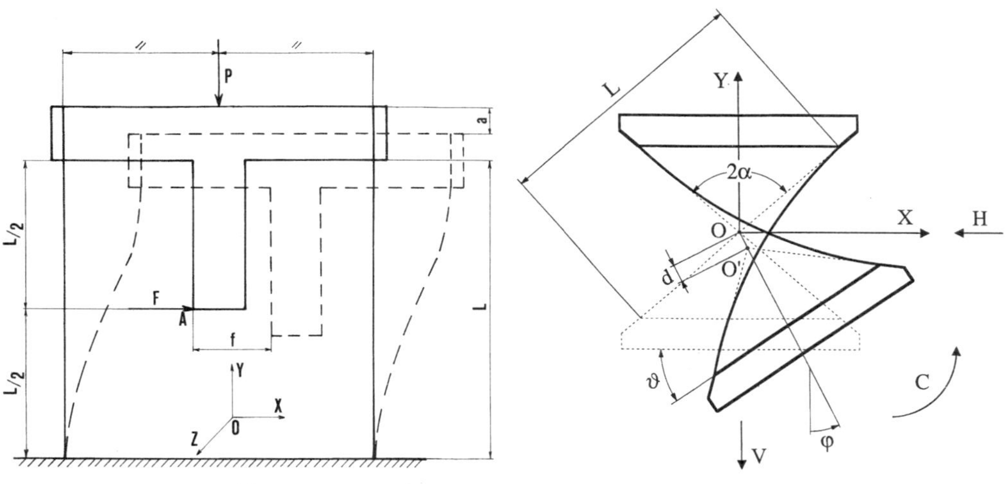


**Figure 2.** Parallel spring translator (a) and cross-spring pivot (b).

Usually all leaf spring based compliant mechanisms have working ranges such that parasitic displacements can arise. For instance, in the case of parallel spring translators (Fig. 2a) the main displacement $f$ along the $x$-axis will be accompanied by the parasitic motion $a$ along the $y$-axis, while in the case of cross-spring pivots (Fig. 2b) the main rotation movement $\vartheta$ will be coupled to a parasitic motion OO' of the geometric centre O of the pivot characterized by the respective magnitude $d$ and phase $\varphi$.

In order to evaluate these effects, models encompassing the non-linear mechanical behaviour of the beams must be adopted. All the considered configurations can therefore be referred to the general case of a cantilever beam loaded at the free end with couples (an initial curvature of the beam itself – e.g. due to residual stresses from the microfabrication process, can also be considered as a couple) and inclined forces (Fig. 3).

This case can be appropriately studied using the "Elastica" approach originally developed only for the case of straight slender and inextensible cantilevers loaded axially (Timoshenko and Gere, 1961) which was recently extended to the generalized case shown on Fig. 3 (De Bona and Zelenika, 1997). A curvilinear coordinate system whose orientation is determined by the inclination of the force is thus introduced. The differential equation of the deflection curve is then given by:

$$M_A = M - Fy = EI\left(\frac{d\theta}{ds} - \frac{1}{r}\right) \tag{2-1}$$

where the exact expression for the curvature of the leaf spring is used.

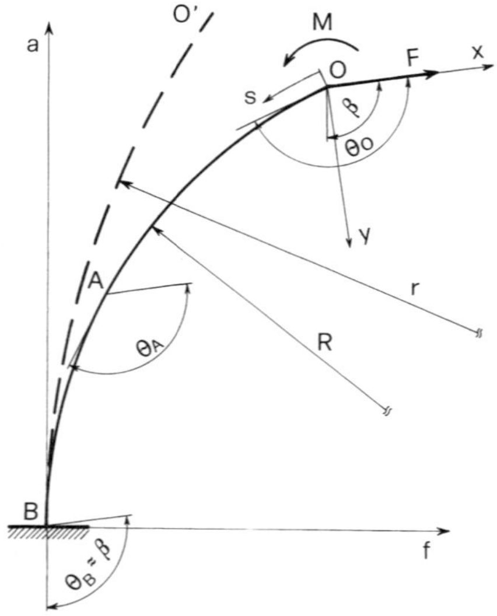


**Figure 3.** Cantilever beam with the general case of loading.

According to the procedure described extensively by De Bona and Zelenika (1997), the global Cartesian coordinates $a$ and $f$ of a generic point A of the elastic line of the spring-strip will then be determined from:

$$\frac{a_A}{L} = \frac{x_B}{L}\cos\beta + \frac{y_B}{L}\sin\beta - \frac{s}{L}\left(\frac{x_A}{s}\cos\beta + \frac{y_A}{s}\sin\beta\right) \tag{2-2}$$

$$\frac{f_A}{L} = -\frac{x_B}{L}\sin\beta + \frac{y_B}{L}\cos\beta + \frac{s}{L}\left(\frac{x_A}{s}\sin\beta - \frac{y_A}{s}\cos\beta\right) \tag{2-3}$$

The following notation is adopted in these expressions:
$L$ – total length of the leaf spring
$s$ – length of a section of the beam
$x_B$, $y_B$ – Cartesian coordinates of point B of beam fixation in the reference frame linked to the inclination of force $F$
$\beta$ - slope angle of force $F$ with respect to axis $a$
and

$$\frac{x_B}{L} = 2\frac{E(k,\varphi_O) - E(k,\varphi_B)}{F(k,\varphi_O) - F(k,\varphi_B)} - 1 \tag{2-4}$$

$$\frac{y_B}{L} = \frac{2k(\cos\varphi_B - \cos\varphi_O)}{F(k,\varphi_O) - F(k,\varphi_B)} \tag{2-5}$$

In the last expressions $F(k, \varphi_i)$ and $E(k, \varphi_i)$ denote respectively the elliptic integrals of the first and second kind, $k$ is the parameter of integration, and $\varphi$ the amplitude of the elliptic integrals (with indexes B and O denoting the respective points on the leaf spring). These parameters are defined with:

$$F(k,\varphi_i) = \int_0^{\varphi_i} \frac{d\varphi}{\sqrt{(1 - k^2 \sin^2\varphi)}} \tag{2-6}$$

$$E(k,\varphi_i) = \int_0^{\varphi_i} \sqrt{(1 - k^2 \sin^2\varphi)}\, d\varphi \tag{2-7}$$

$$k = \sin(\alpha/2) \tag{2-8}$$

$$\sin\varphi = \frac{\sin(\theta/2)}{k} \tag{2-9}$$

where:

$$\cos\alpha = \cos\theta_O - \frac{1}{2C^2}\left(\frac{M}{EI} + \frac{1}{r}\right)^2 \tag{2-10}$$

$\theta$ is the slope angle of the elastic curve with respect to force $F$, and the auxiliary parameter $C$ is:

$$C = \sqrt{\frac{F}{EI}} \tag{2-11}$$

The described approach allows then even the post-buckling behaviour of the leaf springs to be modelled.

Due to the presence of elliptic integrals the outlined model becomes transcendental and it is possible to solve it only following an iterative procedure with successive approximations. Since therefore the elliptic integrals must be repetitively calculated, it becomes suitable to use the respective approximate algorithms based on interpolating procedures or other approximate solutions reported in literature (see e.g. Abramowitz and Stegun, 1970 and Gradshteyn and Ryzhik, 1980). De Bona and Zelenika (1997) have, however, shown that in the case when $\varphi_i = \pi/2$, i.e. for the complete elliptic integrals, only the arithmetic-geometric mean method permits correct values of the integrals to be obtained. In the case of the incomplete elliptic integrals of the first kind, only the ascending Landen transformation method gives appropriate results in the whole interval of integration. In the case of the incomplete elliptic integrals of the second kind, none of the approximate methods is appropriate (marked errors are always generated by their application in a part of the domain of values of the integration parameters), and thus a numerical method of integration has to be used (Fig. 4).

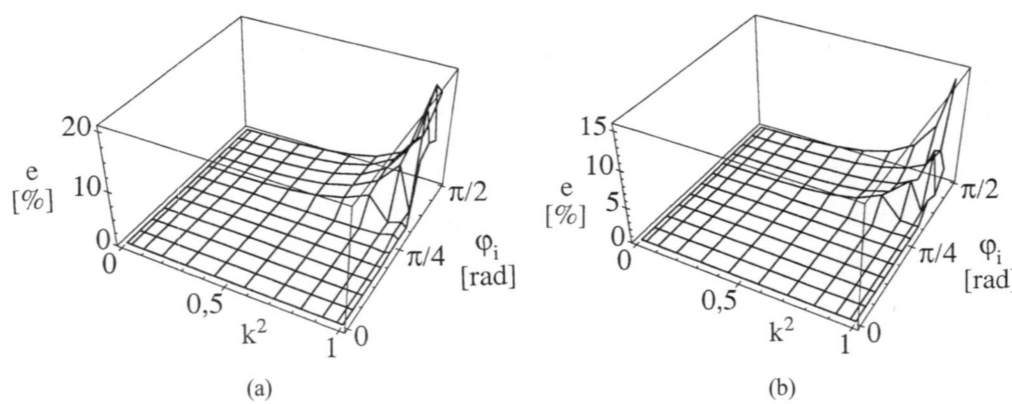


**Figure 4.** Relative error introduced by the arithmetic-geometric mean method in the calculation of the incomplete elliptic integral of the first (a) and second (b) kind.

The experimental verification of the above analytical approach performed by De Bona and Zelenika (1993) and Zelenika and De Bona (2002) by using a high precision laser interferometric measurement technique, showed that the method allows a level of accuracy higher than the interval of uncertainty of the experimental measurements (on the order of $10^{-4}$ of the performed primary motion) to be achieved.

In order to minimize parasitic displacements, Henein (2001), Jones (1988) and Smith and Chetwynd (1992) suggested "compensated" design configurations used recently also in microsystems technology e.g. by Jaecklin et al. (1992) for memory mass storage devices or by Bhushan (2004) for friction measurement equipment. In this case (Fig. 5) two simple translators are mounted onto each other thus causing the overall parasitic motion (sum of equal and opposite contributions from the two translators) to vanish. In reality this will be true only if the lateral (in plane and orthogonal to direction of motion) loads are negligible. If this is not the case, De Bona and Zelenika (1993) showed that the above outlined approach allows to "tune" the design parameters of the two single translators so as to reduce the resulting parasitic motion to negligible levels.

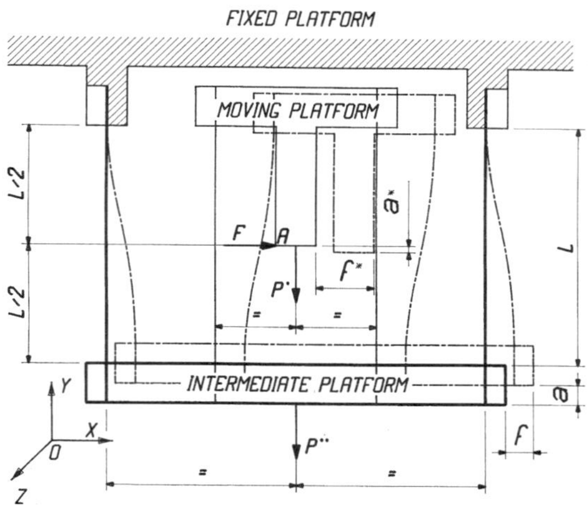


**Figure 5.** Compensated parallel spring translator.

In the case when the overall mechanisms becomes complex, or the loads on the leaf springs are different from those of Fig. 3, the Elastica-type approach cannot be used and a numerical approach generally based on a finite element method (FEM) has to be preferred. In this case, due to the fact that a non-linear large displacement analysis has to be performed, the computational effort by using the commercially available codes could be intensive. For this reason Munteanu et al. (1996) have proposed the usage of a special finite element type (SFET) constituted by 3 nodes, each with a single angular DOF. As the SFET does not require an incremental approach to be followed, fast and accurate results can be achieved with very few elements. A comparison of the results obtained by using SFET with those obtained by using Elastica as well as experimental results showed an excellent agreement.

# 3 Compliant Mechanisms Employing Flexural Hinges

Another widely used compliant mechanism configuration is that based on flexural hinges, where the compliance is localised in determined spots of the considered device (see e.g. Lobontiu, 2003, Lobontiu and Garcia, 2005, Smith, 2000, Smith and Chetwynd, 1992). A traditional design solution is obtained by machining a circular cut-out on a rectangular cross section blank thus obtaining a marked increase of the flexural compliance in the plane of the notch, while retaining the stiffness along the other degrees of freedom (Paros and Weisbord, 1965). If the movable rigid member connected to the notch is much longer than the notch itself, the resulting motion can be determined by considering simple rigid body kinematics models.

Up to recently the choice of the notch shapes for flexural hinges was determined by the available production technologies. In fact, the notches were mainly produced by conventional rotating machine tools and therefore limited to circular shapes. The availability of high-precision milling and especially wire electro-discharge machining (EDM), as well as micro- and meso-manufacturing technologies has allowed these limitations to be overcome. In fact, the most commonly used microfabrication processes (bulk and surface micromachining and LIGA – Fig. 6) are based on lithography (Madou, 2002), a process that permits prismatic microstructures, characterised by complex in-plane shapes, to be easily obtained. It should be noted that micro-systems production techniques have been developed and optimised for in-plane structures, and therefore the geometrical complexity does not imply any additional effort, while on the other hand the high-end production technologies which could be used to achieve the same results in the meso- and macro-fields imply technological difficulties and significant costs.

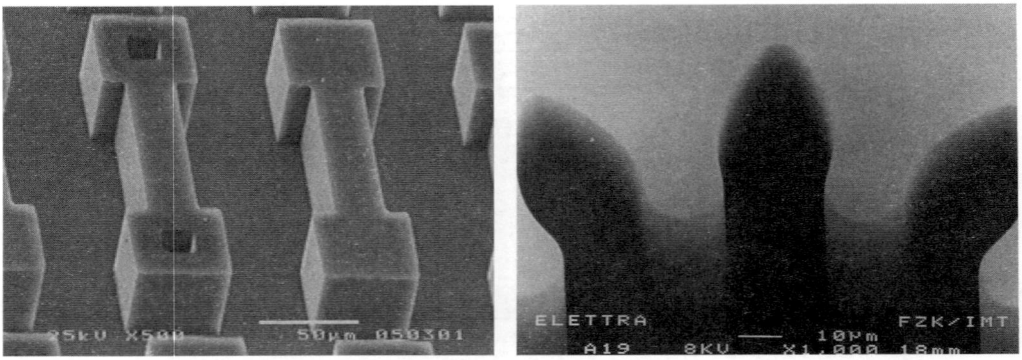


**Figure 6.** Complex in-plane geometries obtained by using deep X-ray lithography.

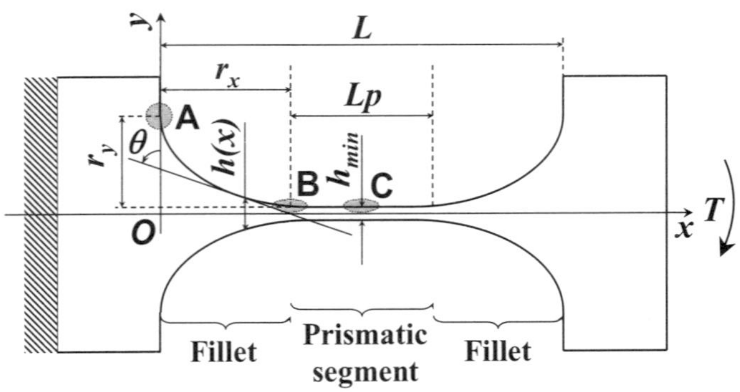


**Figure 7.** Example of a flexural hinge with a non-conventional shape.

In the case of micromechanisms, the shape of the notches (Fig. 7) can thus be chosen based on the design requirements for the specific application. This implies, however, that the mechanical model of the flexural hinge is required. This model has to comprise the accurate evaluation of the compliance, the stresses and the eventual parasitic motions of mechanisms based on the hinges.

The evaluation of compliances is generally performed by means of the Euler-Bernoulli beam models such as those used by Paros and Weisbord (1965). Even if strongly simplified, this approach gives reliable results also for non conventional hinge shapes (parabolic, hyperbolic, elliptical, inverse parabolic and secant profiles) as was shown in Lobontiu (2003) where a numerical validation of this approach by using FEM is performed.

The main drawback of flexural hinges is that a high stress-concentration effect is present which limits their fatigue strength. It seems therefore appropriate to consider already in the hinge design phase not only stiffness but also strength requirements. For this purpose it is often necessary to refer to a numerical model based on FEM. In fact, parametric solutions obtained empirically or analytically in closed form are given in Peterson (1974) only for simple geometries.

What is more, in the case of a conventional rotational joint the relative rotation occurs about an axis passing through its geometrical centre. In the case of flexural hinges of other shapes, even for small displacements, the actual centre of rotation does not coincide with the geometrical centre (the deflection is here distributed along the hinge), thus producing a parasitic motion (Fig. 8). In most cases this parasitic motion can be evaluated accurately only by following a numerical approach, which becomes absolutely necessary in those cases where high flexural rotation angles are aimed at and non-linear effects due to large deflections must be considered. Thus the analytical solution given in Lobontiu (2003) allows only the displacements of the geometrical centre of the there considered hinge shapes to be computed.

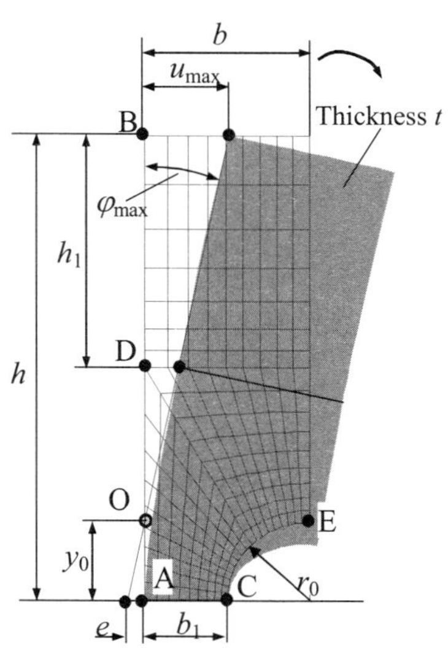


**Figure 8.** Parasitic deflection of a flexural hinge with semi-circular notches.

Fig. 8 shows the case of a conventional notch loaded with a pure couple where the conditions of symmetry and anti-symmetry enable to model numerically only a quarter of the hinge. The deflection of the notch induces a rotation of the movable member around point O instead of around the geometric centre A. For small displacements, point O remains at a fixed position defined by the coordinate $y_0$ which will be obviously different for each hinge shape. De Bona and Munteanu (2005) have, however, shown that, if geometrical non-linearities have to be taken into account, a vertical displacement of point B of the movable rigid block will result, but this displacement is influenced very little by the shape of the hinge.

In any case the adaptation of the shape of the notches to the considered application makes necessary its optimisation. For this purpose a possible approach is that described by De Bona and Munteanu (2005) where a parametric FEM model is coupled to an optimization algorithm. The objective function to be minimised in this case is the flexural stiffness, while the constraints are given by strength (maximal allowable stress) and kinematical (maximal admissible parasitic motion) requirements. The design variables to be optimised are then the geometric parameters defining the notch shape (in De Bona and Munteanu (2005) these are defined via suitable spline functions). In Fig. 9 are shown the results obtained by employing this approach while maintaining constant the value of the strength constraint and varying the kinematical constraint; it can be clearly seen that by allowing the parasitic shift to grow, the shape of the notch tends to become elongated.

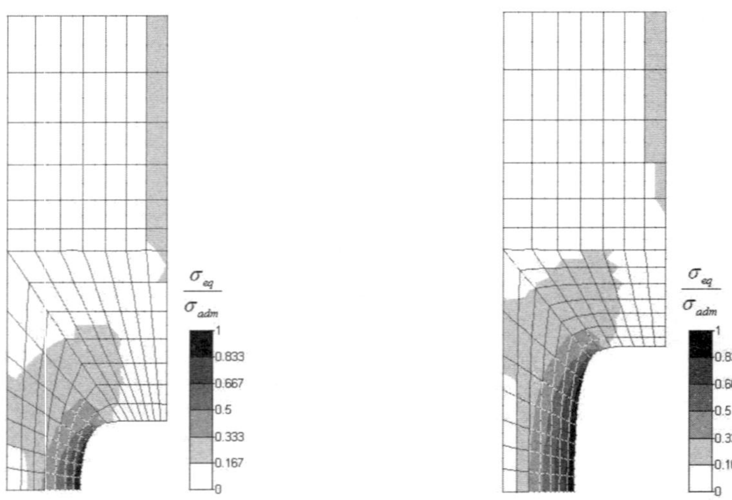


(a) (b)

**Figure 9.** Normalised stress distributions for optimised notch shapes for a small (a) and a larger (b) value of the admissible parasitic shift $y_0$.

In fact, Zelenika et al. (2004) compared the ideal compliant configuration of a leaf spring and the conventional right circular notch shape with optimised notch fillet (transitions between the bulk material and the hinge – Fig. 7) circular, elliptic and parabolic shapes, as well as with "streamline" fillet shapes (Peterson, 1974) obtained in classical mechanics via stress concentration minimisation criteria (Fig. 10).

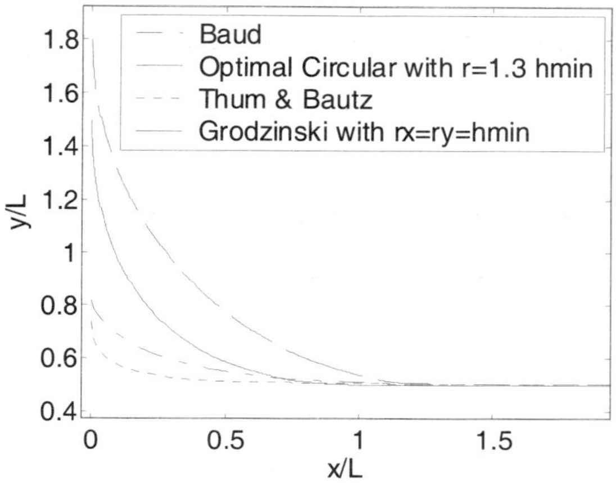


**Figure 10.** Shapes compared to the leaf spring and the circular notch in terms of stress and parasitic shift.

It could thus be clearly shown that a compliance increase can be obtained only at the expense of an increase of the parasitic shift. Depending on the particular application, the optimal shape to be used will thus be based on a trade-off between stress (strength constraint) and parasitic shift (kinematical constraint). The streamline shapes are then to be preferentially used if the main concern is stress minimisation, while the optimised circular and elliptic shapes provide a good compromise if aiming at a parasitic shift minimisation with still far smaller stresses than those of conventional circular hinges (Fig. 11).

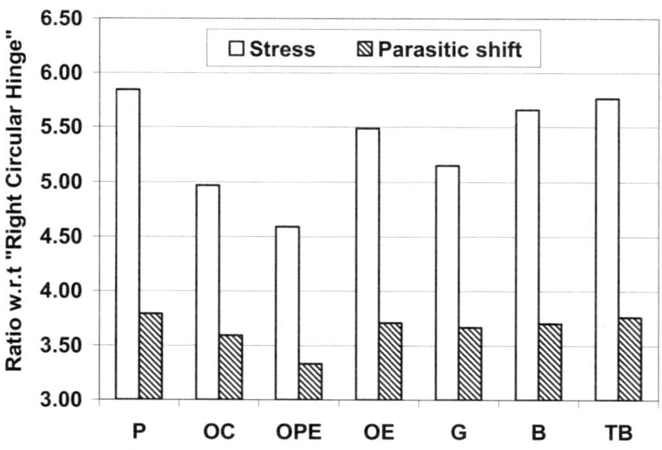


**Figure 11.** Ratio of the normalised stress of the right circular (RC) hinge vs. that of the other hinges (white bars) and ratio of the parasitic shift of the various hinges vs. that of the RC hinge (hachured bars). P – prismatic beam (leaf spring), OC – optimised circular hinge, OPE & OE – optimised elliptic hinges, G, B, TB – streamline hinges

# 4 Continuum Structures With Distributed Compliance

Two approaches to synthesise compliant mechanisms are possible (Nishiwaki et al, 1998): kinematical synthesis and continuum synthesis approach.

In the kinematical synthesis approach the compliant mechanism is obtained simply by introducing lumped compliance (leaf springs or flexural notches) in a traditional rigid link configuration.

In the case of the continuum synthesis approach, a topology optimization based design is followed (Howell, 2001). On Fig. 12 are thus represented different topologies in the two columns (a)-(c) and (b)-(d) and different shapes for the same topology in the two rows ((a)-(b) and (c)-(d)).

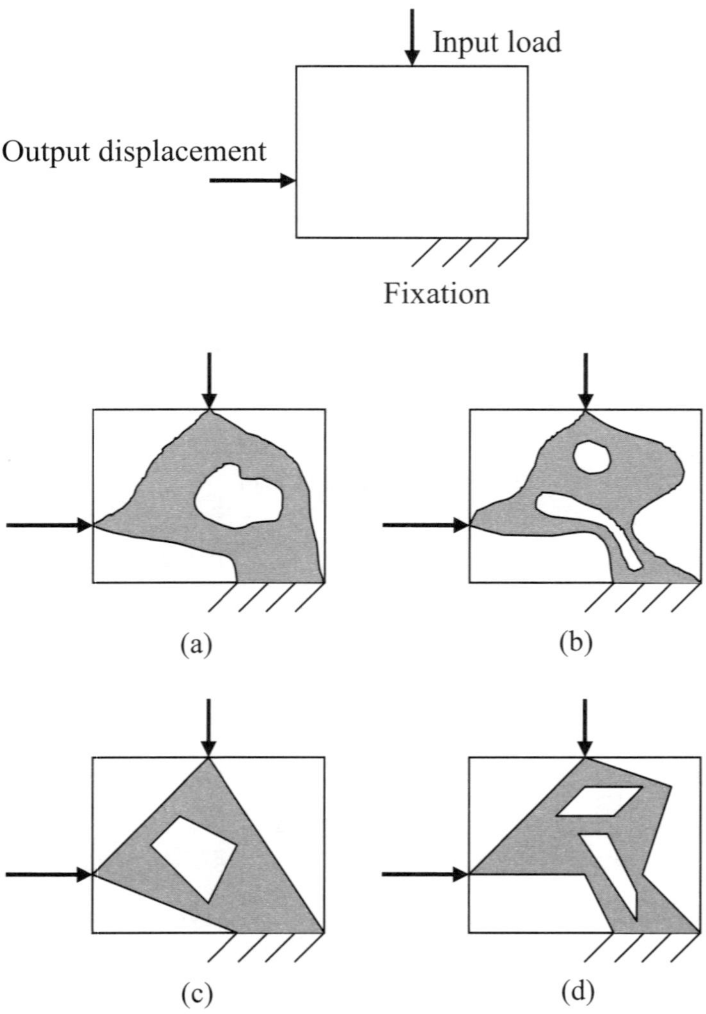


**Figure 12.** Continuum synthesis approach.

Hetrick and Kota (1999) describe thus a simple topology optimisation technique called "ground structure parameterisation". The design space is here defined by a mesh of truss elements that are removed as the analysis determines that they are unnecessary (Fig. 13).

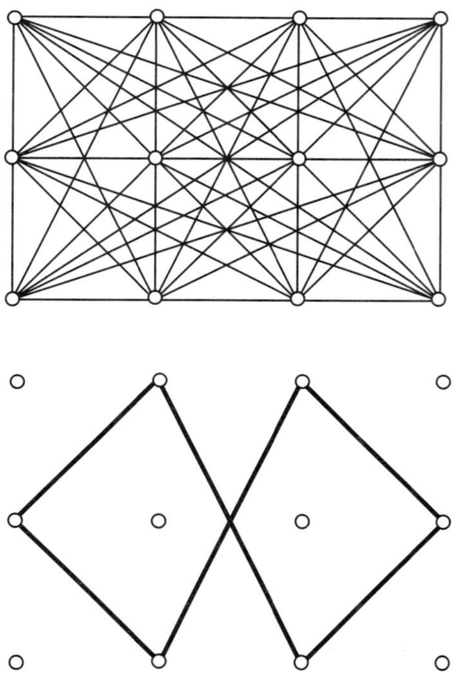

**Figure 13.** Topology optimisation based on truss elements removal.

In the "continuous material density parameterisation" approach, the design space is a region of material modelled as a grid (Fig. 14). As described by Eschenauer and Olhoff (2001), the applied analysis varies then the density of the material in each cell of the grid, leaving a rough image of the compliant mechanism that is finally refined performing a classical shape optimization. This approach is widely used in the design of compliant micromechanisms (see e.g. Ananthasuresh and Kota, 1995 and Kota et al., 2001).

The above approach was derived from a more general method called "homogenization method". In this case the material inside the design domain is treated as a composite made of a microstructure consisting of a material and a void. For each element in the discretised design domain, its void is defined using three parameters $a$, $b$ and $\theta$ (Fig. 15). The optimization algorithm determines the value of these parameters for each element of the grid.

These approaches, even if quite promising, are still in the refinement stage and at present no commercial tools based on them are available. Moreover, these methods seem not to be able to take into account very accurately during the optimization phase stress concentrations nor the parasitic motions of the mechanism.

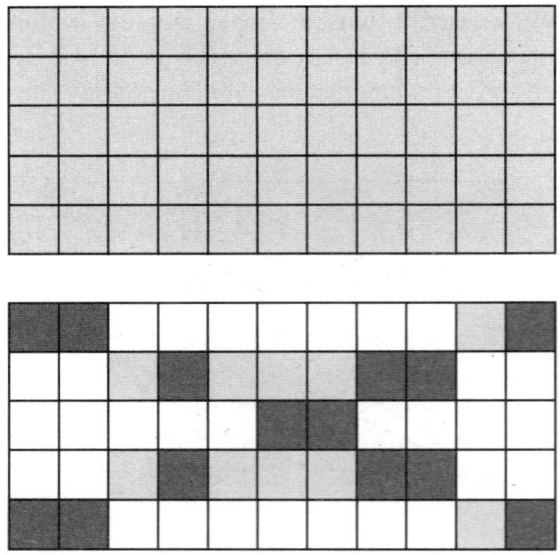

**Figure 14.** Continuous material density topology optimisation.

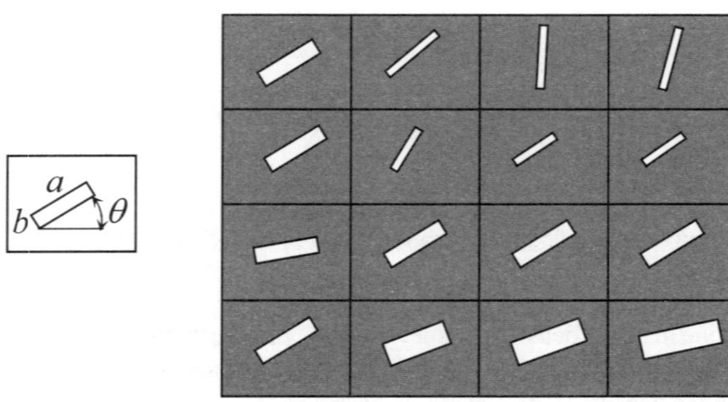


**Figure 15.** Homogenisation method approach.

## References

Abramowitz, M., and Stegun, I. A., ed. (1970). *Handbook of Mathematical Functions with Formulas*. New York, NY, USA: Dover Publications.

Ananthasuresh, G. K., and Kota, S. (1995). Designing compliant mechanisms. *Mechanical Engineering*, November:93-99.

Bhushan, B., ed. (2004). *Springer Handbook of Nanotechnology*. Berlin, D: Springer.

De Bona, F. and Zelenika, S. (1993). Characterisation of High Precision Parallel Spring Translators. In Ikawa, N., Shimada, S., Moriwaki, T., McKeown, P. A., and Spragg, R. C., eds., *International Progress in Precision Engineering*. Stoneham, MA, USA: Butterworth-Heinemann. 761–772.

De Bona, F., and Zelenika, S. (1997). A generalised Elastica-type approach to the analysis of large displacements of spring-strips. *Journal of Mechanical Engineering Sciences – Proceedings of the Institution of Mechanical Engineers* C211:509-517.

De Bona, F., and Munteanu M. Gh. (2005). Optimized Flexural Hinges for Compliant Micromechanisms. *Analog Integrated Circuits and Signal Processing* 44:163–174.

Eschenauer, H. A., and Olhoff N. (2001). Topology optimization of continuum structures: a review. *Applied Mechanics Review* 54:331-390.

Gradshteyn, I. S., and Ryzhik, I. M. (1980). *Table of Integrals, Series and Products*. San Diego, CA, USA: Academic Press.

Henein, S. (2001). *Conception de guidages flexibles*. Lausanne, CH: Presses Polytechniques et Universitaires Romandes.

Hetrick, J., and Kota, S. (1999). An Energy Formulation for Parametric Size and Shape Optimization of Compliant Mechanisms. *Journal of Mechanical Design* 21(2):229-234.

Howell, L. L. (2001). *Compliant Mechanisms*. New York, NY, USA: John Wiley & Sons.

Jaecklin, V. P., Linder, C., de Rooij, N. F., and Moret, J. M. (1992). Micromechanical comb actuators with low driving voltage. *Journal of Micromechanics and Microengineering* 2:250–255.

Jones, R. V. (1988). Instruments and Experiences – Papers on Measurement and Instrument Design. New York, NY, USA: John Wiley & Sons.

Kota, S., Joo, J., Li, Z., Rodgers, S. M., and Sniegowski, J. (2001). Design of compliant mechanisms: applications to MEMS. *Analog Integrated Circuits and Signal Processing* 29:7-15.

Lobontiu, N. (2003). *Compliant Mechanisms – Design of Flexible Hinges*. Boca Raton, FL, USA: CRC Press.

Lobontiu, N., and Garcia, M. (2005). *Mechanics of Microelectromechanical Systems*. Norwell, MA, USA: Kluwer Academic Publishers.

Madou, M. (2002). *Fundamentals of Microfabrication – The Science of Microfabrication*. Boca Raton, FL, USA: CRC Press.

Munteanu, M. Gh., De Bona, F., and Zelenika, S. (1996). An accurate non-linear analysis of very large displacements of beam systems. In *Proceedings of the International Conference on Material Engineering*. Gallipoli - Lecce, I. 59–66.

Nishiwaki, S., Min, S., Ejima, S., and Kikuchi, N. (1998). Structural optimization considering flexibility. *International Journal of the Japanese Society of Mechanical Engineers*, C41:476-484.

Paros, J. M., and Weisbord, L. (1965). How to design flexures hinges. *Machine Design* November: 151–156.

Peterson, R. E. (1974). *Stress Concentration Factors*. New York, NY, USA: John Wiley & Sons.

Senturia, S. D. (2001). *Microsystem Design*. Norwell, MA, USA: Kluwer Academic Publishers.

Slocum, A. H. (1992). *Precision Machine Design*. Dearborn, MI, USA: Society of Manufacturing Engineers.

Smith, S. T., and Chetwynd, D. G. (1992). *Foundation of Ultraprecision Mechanism Design*. Amsterdam, NL: Gordon & Breach Science Publishers.

Smith, S. T. (2000). *Flexures – Elements of Elastic Mechanisms*. Amsterdam, NL: Gordon & Breach Science Publishers.

Timoshenko, S. P, and Gere, J. M. (1961). *Theory of Elastic Stability*. New York, NY, USA: McGraw-Hill.

Trylinski, W. (1971). *Fine Mechanisms and Precision Instruments – Principles of Design*. Oxford, UK: Pergamon Press.

Yang, G., and Nelson, B. J. (2004). Automated Microassembly. In Hsu T.-R. *MEMS Packaging*. London, UK: INSPEC.

Zelenika, S., and De Bona, F. (2002). Analytical and Experimental Characterisation of High-Precision Flexural Pivots Subjected to Lateral Loads. *Precision Engineering* 26(4):381-388.

Zelenika, S., Henein, S., and Myklebust, L. (2004). Investigation of Optimised Notch Shapes for Flexural Hinges. In *Proceedings of the 3rd International Workshop on the Mechanical Engineering Design of Synchrotron Radiation Equipment and Instrumentation*. Grenoble, F. paper 04-25.

# Micro- and Nano-assembly and Manipulation Techniques for MEMS

Eniko T. Enikov
Department of Aerospace and Mechanical Engineering, University of Arizona, Tucson, AZ, USA

**Abstract** This paper presents a review of the most commonly used techniques for the assembly of micro-systems. Following a brief overview of the dominant forces working at this scale, the operation and design of mechanical, electrostatic, and magnetic micro-grippers is presented. The use of electrostatic forces is further described in the context of nano-assembly, where sub-micron-sized charged spots are used as the anchoring sites for nano-particles.

## 1 Micro- and Nano-assembly: Purpose and Defining Characteristics

Although MEMS devices are usually fabricated via massively parallel photolithographic techniques, sequential assembly is required in some instances. For example, heterogeneous integration for vertical cavity surface emitting lasers (VCSELs) with silicon-based CMOS circuitry requires placement of the laser die onto a silicon substrate containing the electronic circuitry. Further applications include the assembly of dense arrays of high aspect-ratio structures such as IC probe cards or light-emitting diodes in display applications. These assembly and packaging operations are costly and usually constitute the largest portion of the device's total cost. In order to increase the manufacturing throughput and reduce the re-tooling costs, it is desirable to develop flexible assembly schemes, allowing for quick adaptation to various part geometries and configurations. Responding to this need, research in this area has led to the development of micro-assembly techniques and robotic systems attempting to manipulate and assemble sub-millimeter-sized parts into functional devices (Vikram et al., 2000).

The term *micro-assembly* has emerged to describe the assembly of components with sizes below 1 mm, and nano-assembly refers to manipulation of parts with sizes below 100 nm. Micro- and nano-assembly can be *sequential*, when parts are assembled one-at-a-time, or *parallel*, when multiple components are manipulated and assembled simultaneously. Sequential assembly is more common in conventional (macro-scale) manufacturing, but parallel assembly is desired for MEMS devices, since they are produced photolithographically in a massively parallel fashion. Micro-assembly is most frequently needed when the parts cannot be fabricated on the same substrate due to process or material constraints. This is the case for optical devices such as laser diodes and detectors, which are fabricated on GaAs, GaN, and other wide-band-gap materials. For example, semiconductor laser diodes with a 300 $\mu$m footprint are commonly assembled into CD-ROM drives (Hara et al., 1993).

Parallel micro-assembly can be *deterministic* if the relationship between the parts and their destination is known in advance, or *stochastic* when this relationship is random. Flip-chip wafer-to-wafer transfer (Cohn and Howe, 2000; Singh et al., 1999) is an example of deterministic assembly. Conversely, the process demonstrated by Jacobs et al. (2002) for the assembly of light-emitting diodes in a liquid bath represents a stochastic assembly.

## 2 Characteristic Features of Micro- and Nano-assembly

The positioning and manipulation of small components has its own peculiarities stemming from the increased role of surface forces. As the size of the part is reduced, its surface area-to-volume ratio increases, leading to effects such as sticking and jumping during handling. Below a certain critical dimension, it is not possible to simply release these parts by using gravity. These effects could be understood better by considering the main forces acting on a spherical particle with radius $r$ in close proximity to a planar surface (see Fig. 1). The body force, e.g. gravity, is given by

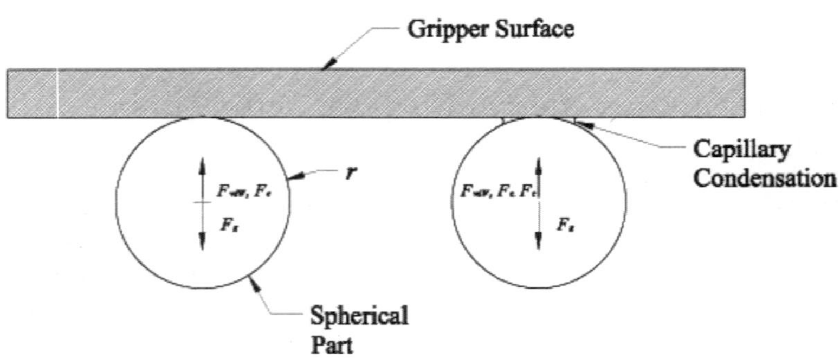


**Figure 1.** Forces acting on a sphere of radius $r$ in close proximity to a surface.

$$F_g = \frac{4}{3}\pi\rho g r^3, \tag{2.1}$$

where $\rho$ is the density of the part and $g$ is the Earth's gravitational field. When such a part carries charge $Q$ and is brought in proximity to a conductive plane (e.g. the surface of the gripper), the image charge induced in the gripper results in electrostatic attraction, which can be estimated as

$$F_{elec} = \frac{Q^2}{16\pi\epsilon r^2}, \tag{2.2}$$

where $\epsilon$ is the dielectric constant of the medium surrounding the sphere. The particle charge $Q$ could be due to contact electrification resulting from the contact potential difference between two metals, or it could be a result of the trapped charge stored in a dielectric layer on the surface of the part. In the case of contact electrification, the surface charge is accumulated through tunneling as the two spheres are brought close to one another. The resulting surface charge density, $\sigma$, can be estimated by $\sigma = \epsilon_0 \Phi / z_0$, where $\Phi \approx 0.5$ V is the metal work function difference and $z_0 \approx 1$ nm is the tunneling distance (Lowell and Rose-Innes, 1980). These values result in charge densities on the order of 4 $mC/m^2$. Similarly, charges can accumulate in dielectric layers and their interfaces. In silicon dioxide, typical values are in the range 0.6 – 1.1 $mC/m^2$ (Nicollian and Brews, 2003). The total charge on the sphere can then be estimated by $Q = 4\pi r^2 \sigma$.

The sphere will also experience van der Waals forces given by (Bowling, 1988)

$$F_{vdW} = \frac{Ar}{6z^2}, \tag{2.3}$$

where $z$ is the distance between the sphere and the plane and $A$ is the Hamaker constant (Israelachvili, 1991), whose value depends on the properties of the two surfaces and is on the order of $A \approx 10^-19$ J for solid surfaces. This force is surprisingly strong. For example, a sphere of radius 1 cm within 0.2 nm from a surface would experience $F_{vdW} = 4$ mN of force, which is easily measurable with standard equipment. Finally, if the particle and the surface are surrounded by a gas phase that can adsorb (condense) on the surface, then liquid meniscus is formed, a phenomenon known as capillary condensation (see Fig. 1(right).) The capillary force is given by the surface tension of the adsorbed film, $\gamma_L$, and the liquid contact angle, $\theta$ (Israelachvili, 1991)

$$F_{cap} = 4\pi r \gamma_L \cos\theta. \tag{2.4}$$

A plot of the forces listed in Eqs. (2.1-2.4) versus the radius of the sphere is shown in Figure 2. As evident, the electrostatic and van der Waals forces dominate gravitational

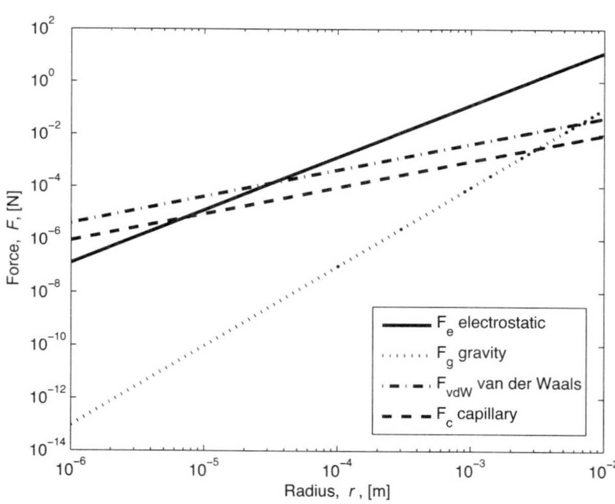


**Figure 2.** Relative magnitudes of forces acting on a sphere vs. its radius: gravitational, electrostatic, van-der-Waals, and capillary

forces for parts with dimensions below 2 – 3 mm. The immediate consequence of this is that small parts cannot be placed down by simply releasing them. Rather, active releasing techniques are needed to overcome the surface adhesion forces. Another consequence of the dominant surface forces is the inaccuracy of placement using conventional mechanical tweezers. Figure 3 illustrates this point. Numerous techniques for manipulating micro-

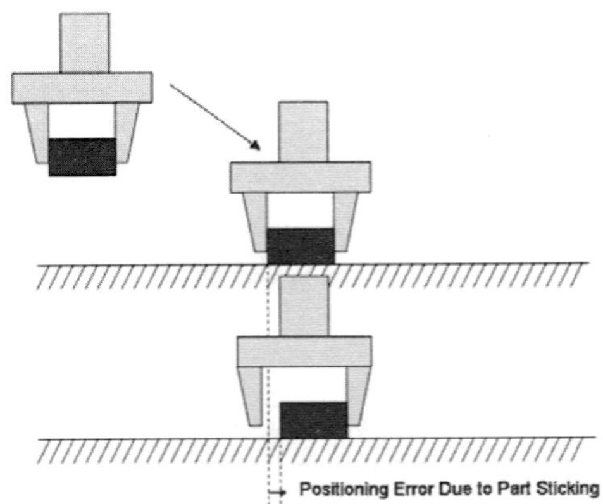


**Figure 3.** Part placement error due to surface adhesion

parts have been explored in order to overcome the van der Waals and capillary forces. The control of the adhesion can be achieved through sharp protrusions designed to reduce the influence of van der Waals forces. Figure 4(left) illustrates the pick and release of a sphere using this concept. The success of this technique requires careful motion planning designed to roll the sphere through a sharp edge, where the attractive van der Waals forces are reduced (Miyazaki and Zato, 1996; Arai et al., 1996; Haliyo et al., 2001). Heating has also been used for controlling the surface forces. Arai and Fukuda (1997), for example, have developed a gripper containing an array of heated voids on its surface, as illustrated in Figure 4(right). Thermal expansion/contraction of the trapped gas is used to release/attach the grasped part. Repulsive forces generated by eddy currents were explored by Feddema et al. (2002) in an effort to release electrostatically clamped metal parts. While some of these methods produced promising results, the control of surface forces remains one of the most challenging problems in micro-assembly.

## 3 Micro-assembly Techniques and Devices

### 3.1 Mechanical Grippers

Mechanical gripping is the oldest and most common assembly technique. It is not surprising, therefore, that MEMS technology has also been successfully applied to mechanical gripping. Micro-mechanical tweezers have been developed using polycrystalline silicon (Kim et al., 1990; Keller and Howe, 1995), shape memory alloy (SMA) (Kohl et al., 2002; Bellouard et al., 1998; Hesselbach et al., 1997), electroplated metals (Car-

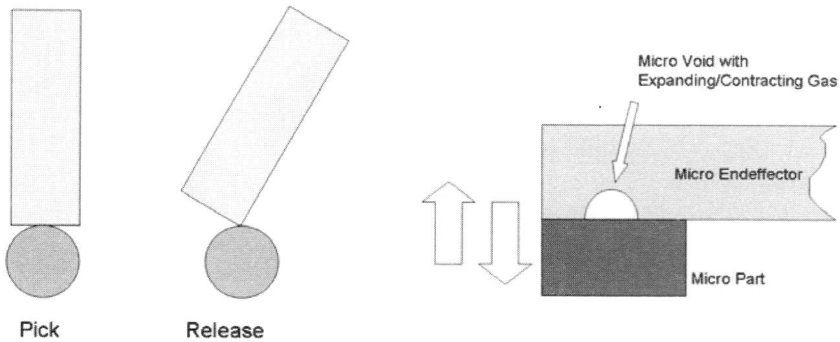


**Figure 4.** Pick up and release using van der Waals force-based gripper [left, (Miyazaki and Zato, 1996)] and using gas expansion/contraction [right, Arai and Fukuda (1997)].

rozza et al., 1998), and piezoceramics (Breguet et al., 1997). Figure 5 shows an example of two micro-tweezers overhanging the edge of a Si chip. Folded-beam thermal actuators (Enikov and Lazarov, 2005) are used to close the jaws of the tweezers. This actuator

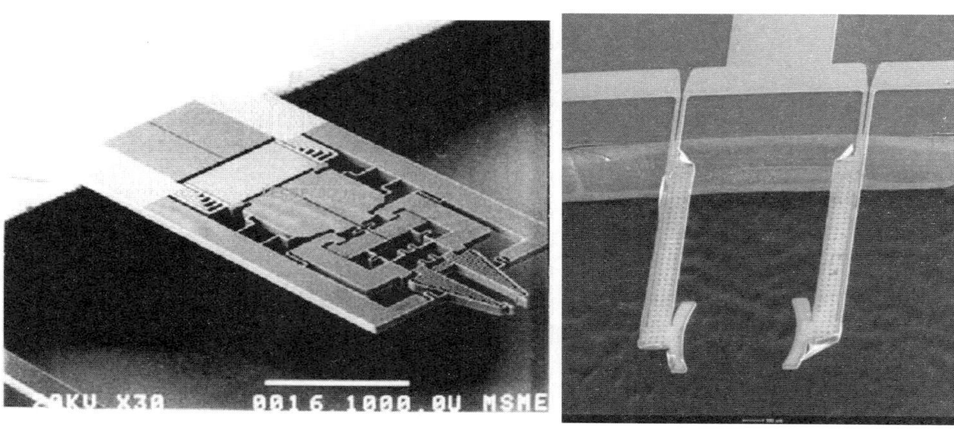

**Figure 5.** Thermally actuated tweezers: (left) polycrystalline silicon (Keller and Howe, 1995), reproduced with permission from MEMSPI.com; (right) electroplated Ni gripper.

type is quite popular due to the simplicity of its design and operation. To produce a gripping action, current is passed along the metal beams, resulting in a non-uniform temperature distribution. The resulting deformation is towards the wider (inner) arm. When properly designed, 1.3-mm-long beams can easily produce displacements of up to 80 $\mu$m (Enikov and Lazarov, 2005). A detailed description of the design of these actuators is provided in an earlier chapter of this book.

### 3.2 Vacuum grippers

Vacuum gripping is a traditional method used for placing surface-mounted components in IC industry. It requires smooth surfaces and relatively large gripping area to produce sufficient gripping force. Vacuum holding is especially useful in the manipulation of biological cells, for example during in-vitro fertilization (IVF) of egg cells (Kamura and Yanagimachi, 1995) (see

Fig. 6. Using a pair of glass pipettes it is possible to hold an egg cell with one of them, while injecting it with a DNA material or spermatozoa via the second pipette.

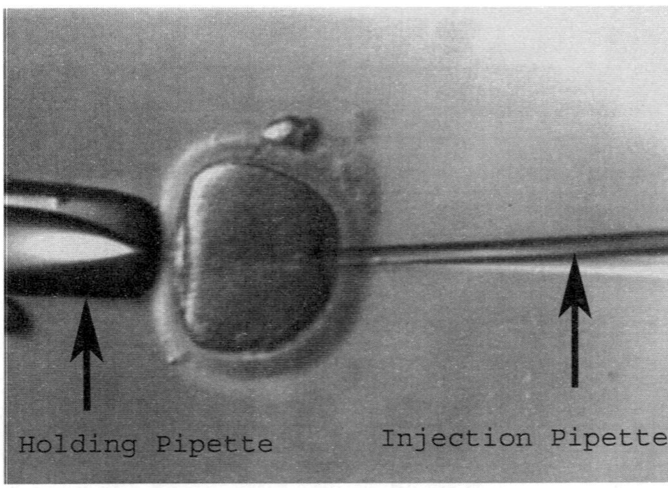


**Figure 6.** Mouse oocyte (egg cell) during intracytoplasmic sperm injection (ICSI). Photo courtesy of Dr. Kenneth Roberts, University of Minnesota.

### 3.3 Electrostatic Grippers

Due to its predictable nature, compatibility with vacuum equipment, and simple application, the electrostatic attachment has also been heavily studied (Gengenbach and Boole, 2000; Feddema et al., 2002; Enikov and Lazarov, 2003; Hesselbach et al., 2003). Numerous variants of the electrostatic attachment are possible, as illustrated in Figure 7. Depending on the number of electrodes, monopolar, bipolar, or multi-polar clamps are used. All of these utilize electrostatic forces of attraction. Depending on the materials and electrode geometry, the resulting attachment forces can vary significantly. While an attractive force exists for both conductive and dielectric objects, the force of attraction of the dielectrics is proportional to their polarizability (dielectric constant), resulting in much weaker attractive forces. The surrounding atmosphere also has a pronounced effect on the gripping force. For example, increasing the relative humidity above 65 %RH resulted in a complete inability to pick up glass micro-spheres (Hesselbach et al., 2003). The utility of an electrostatic attachment for conductive materials is, however, industrially proven. Commercial devices using this mechanism are already in use for

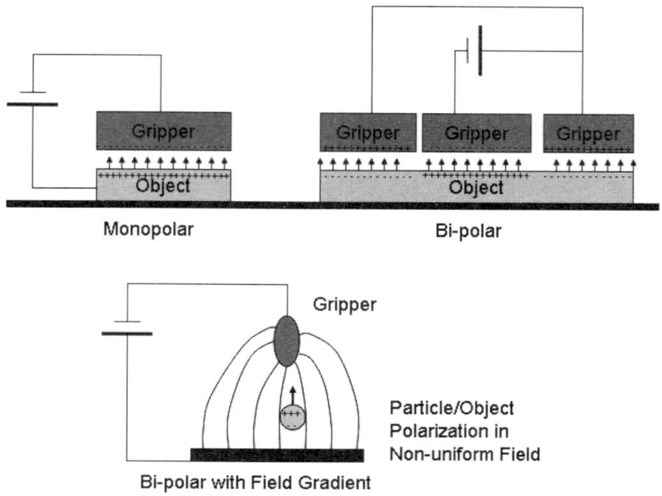


**Figure 7.** Electrostatic gripper: unipolar, bi-polar, and field-gradient based.

clamping silicon wafers during plasma processing and during extreme ultraviolet light lithography (EUVL), for example (Mikkelson et al., 2004).

The design of electrostatic grippers is quite straightforward. A common model of a bi-polar gripper is shown in Figure 8. The surface of the gripper is coated with a dielectric layer to allow manipulation of conductive parts (Enikov et al., 2005). The

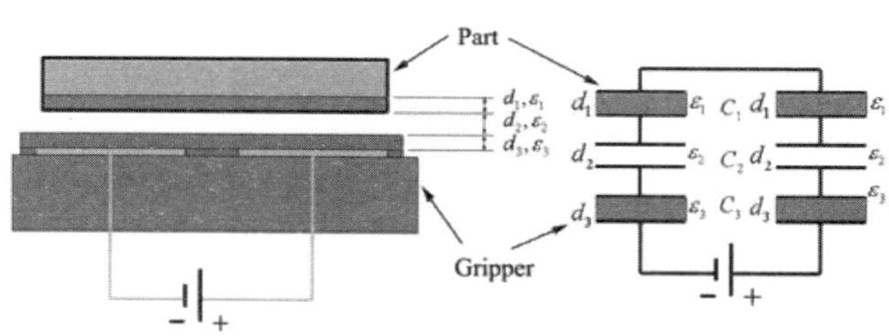


**Figure 8.** Bipolar electrostatic gripper with an air gap and its equivalent circuit

holding force is a function of the total capacitance and can be obtained from the total electrostatic energy of the system $E$ by differentiation

$$F_e = -\frac{\partial E}{\partial d}, \qquad E = \frac{CV^2}{2}, \tag{3.1}$$

where $C$ is the total capacitance of the two equal capacitors in series, $V$ is the total voltage applied between the two electrodes, and $d$ is the gap between the electrodes.

The total capacitance can be estimated (neglecting the fringe fields) from the following equations:

$$\frac{1}{C} = 2\left(\frac{1}{C_1} + \frac{1}{C_2} + \frac{1}{C_3}\right); \qquad C_i = \epsilon_0\epsilon_i\frac{S}{d_i}, i = 1, 2, 3; \quad S = \eta\frac{A}{2}, \tag{3.2}$$

where $A$ is the area of the assembled part (upper electrode), $\eta$ is the filling ratio, $d_2$ is the air gap thickness, $\epsilon_2$ is the air dielectric constant, $d_3$ is the thickness of the dielectric layer on the part (this could be a $SiO_2$ layer on the Si part), and $\epsilon_3$ is the dielectric constant of the dielectric layer. The gripper developed by Enikov and Lazarov (2003) has 10-$\mu$m-wide electrodes at 10-$\mu$m spacing (filling ratio of $\eta = 0.5$ ). Equations (3.1)-(3.2) yield

$$C = \epsilon_0\eta\frac{A}{4\left(\frac{d_1}{\epsilon_1} + \frac{d_2}{\epsilon_2} + \frac{d_3}{\epsilon_3}\right)}, \tag{3.3}$$

with a resulting clamping force of

$$F_e = \epsilon_0\epsilon_2\eta\frac{AV^2}{8d_2^2}\alpha; \qquad \alpha = \frac{1}{\left(1 + \frac{d_1}{d_2}\frac{\epsilon_2}{\epsilon_1} + \frac{d_3}{d_2}\frac{\epsilon_2}{\epsilon_3}\right)^2}. \tag{3.4}$$

This three-layer model can be used to predict the force in the case of a single dielectric layer $(d_1, d_3 = 0)$ between the electrostatic gripper and the conductive part. If the dielectric layer has thickness $d$ and dielectric constant $\epsilon$, Eq. (3.4) reduces to

$$F_e = \epsilon_0\epsilon\eta\frac{AV^2}{8d^2}. \tag{3.5}$$

In order to increase the clamping force of the gripper, one can use a dielectric with a higher dielectric constant $\epsilon$, since the force is directly proportional to the dielectric constant of the gripper coating. Another option is to increase the filling ratio $\eta$, which is limited by the potential dielectric breakdown between neighboring electrodes. In the case of an air gap between the part and the gripper, Eq. (3.4) shows that even if the dielectric constant of the gripper coating is very high ($\epsilon_1 \gg 1$), the force in the system is limited by the air gap, which has a low dielectric constant $\epsilon_2 \approx 1$. For this reason, it is essential to have good planarity between the gripper and the part, and to avoid particulate contamination of the surface.

Using the experimental device shown in Figure 9, Enikov and Lazarov (2003) measured clamping forces using electroplated Ni parts. Because the force could only be applied to the tip of the part, a correction factor was needed to account for the non-uniform force distribution. Under these conditions, the measured force, $F_{\text{meas}}$, is given by

$$F_{\text{meas}} = F_e\frac{L_e}{2L_a} = c_{\text{theor}}V^2, \qquad c_{\text{theor}} = \epsilon_0\epsilon_2\eta\frac{AL_e\alpha}{16d_2^2L_a}, \tag{3.6}$$

where $L_a$ is the arm of the applied force and $L_e$ is the clamped length of the part. Figure 10 shows the result of normal force measurements at various clamping potentials. The quadratic fit to the data resulted in $c_{\text{exper}} = 5.7 \times 10^{-5}$ mN/V$^2$, which corresponds to an

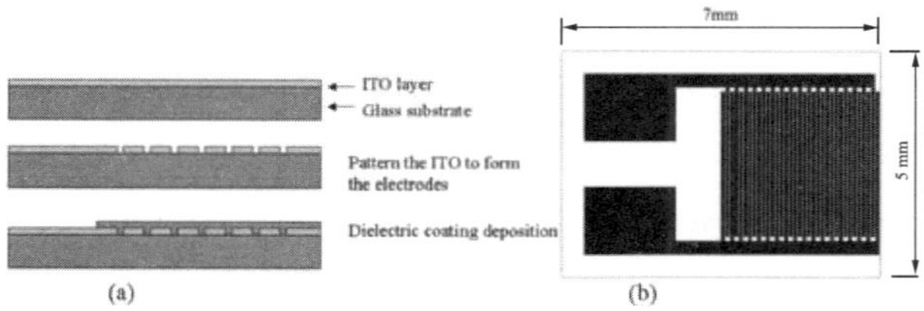


**Figure 9.** (a) Fabrication sequence; (b) photomask used.

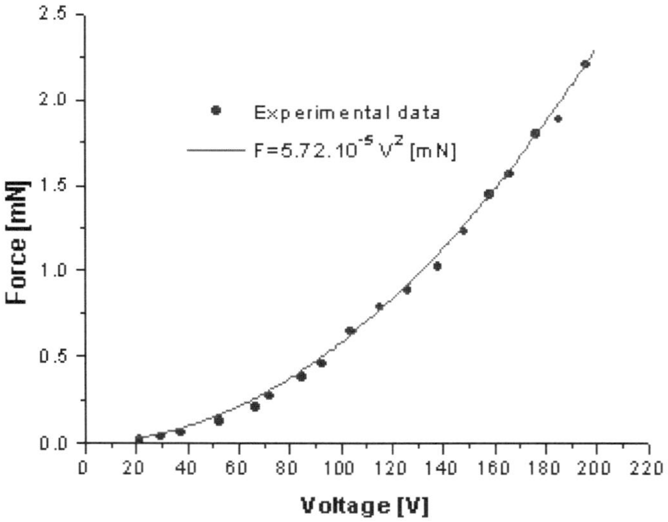


**Figure 10.** Normal force vs. clamping voltage of Ni part.

effective air gap of $d_2 = 1.05\mu$m, consistent with the typical roughness of electroplated Ni surfaces.

An attractive characteristic of the electrostatic grippers is their ability to center (align) the clamped part with respect to a conductive pattern. The centering effect of an electrostatic field can be understood by considering the electrostatic energy stored in a two-plate capacitor, neglecting the fringe fields (see Fig. 11) If the plates are separated laterally or transversally, the total energy stored in the capacitor decreases, i.e. external work is needed to separate the plates and reduce the stored energy. The force then can be calculated from an energy balance, which should include the power supply attached to the capacitor. In this case, the electrostatic force can be determined through the derivative of the total energy (Bao, 2000)

$$F_x = \frac{\partial E}{\partial x}, \qquad F_y = \frac{\partial E}{\partial y}, \qquad F_z = \frac{\partial E}{\partial z}, \tag{3.7}$$

where $E$ is the total electrostatic energy stored in the capacitor and can be calculated through the parallel plate capacitance formula (see Fig. 11 for dimensions and notation)

$$E = \frac{\epsilon_0 (w-x)(v-y)V^2}{2z}. \tag{3.8}$$

From Eqs. (3.7) and (3.8), one can obtain the three forces

$$F_x = -\frac{\epsilon_0 (v-y)V^2}{2z}, \qquad F_y = -\frac{\epsilon_0 (w-x)(v-y)V^2}{2z}, \qquad F_z = -\frac{\epsilon_0 (w-x)(v-y)V^2}{2z^2}. \tag{3.9}$$

Equations (3.9) show that all three forces are directed towards closing the space between the capacitor plates. The lateral forces $F_x$ and $F_y$ are directed opposite to the dis-

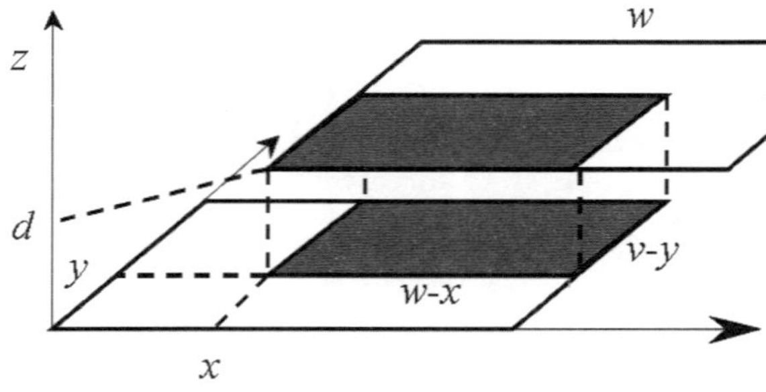


**Figure 11.** Parallel plate capacitor with laterally displaced plates by $x$, $y$, and $d$, respectively.

placements, $x$ and $y$, and tend to align the two plates with respect to one another. The relative magnitudes of the three forces can be evaluated for $x = y = 0$ and $z = d$

$$F_x = -\frac{\epsilon_0 (v)V^2}{2d}, \qquad F_y = -\frac{\epsilon_0 (w)V^2}{2d}, \qquad F_z = -\frac{\epsilon_0 (w)(v)V^2}{2d^2}, \tag{3.10}$$

resulting in $F_x = F_z \times d/w$ and $F_y = F_z \times d/v$. Since, typically, $d/w \approx d/v \approx 1/(100 \ldots 1000)$, one can see that the lateral centering force is 2-3 orders of magnitude smaller than the clamping force. Therefore, in order to achieve self-alignment, the friction coefficient between the part and the gripper should be very small, $\mu \approx 0.01 \ldots 0.001$. Using a 20-kHz ultrasonic field, Bohringer et al. (1998) demonstrated an effective friction coefficient of less than 0.002 for a silicon dioxide part placed over an Au electrode. In addition to the low friction coefficient, the controlled centering of parts requires a part-matching electrode pattern. For example, a concentric electrode design has been used by Oh (1989) and has later optimized by Hesselbach et al. (2003) for centering spherical and cylindrical objects.

When discussing the operation of electrostatic grippers, one has to consider the possibility of electric charge retention. Charge can be trapped at interfaces between conducting and dielectric layers and inside dielectrics (Enikov and Palaria, 2004). This charge can be quantified using a scanning probe microscopy technique known as Kelvin probe microscopy (Jacobs et al., 1999). This is a two-scan technique. During the first scan, the topography of the surface is recorded. The second scan is used to record the electric potential by retracting the probe to a constant height above the surface. Both scans are performed in tapping mode, that is, the cantilever is oscillated with constant amplitude. Since the amplitude of oscillation is a function of the surface/tip distance and the resulting force interactions, any difference between the potentials of the cantilever and the surface will result in a change of the amplitude or phase of oscillation. Thus by applying a bias voltage to the cantilever that is equal to that of the surface, its oscillation amplitude is kept constant. This potential is reported as the surface potential. Figure 12 shows the residual surface potential generated by a trapped charge on the gripper surface 15 minutes after voltage application. The solid white arrow indicates the location of the positive electrodes while the solid black arrow indicates the negative electrodes. The

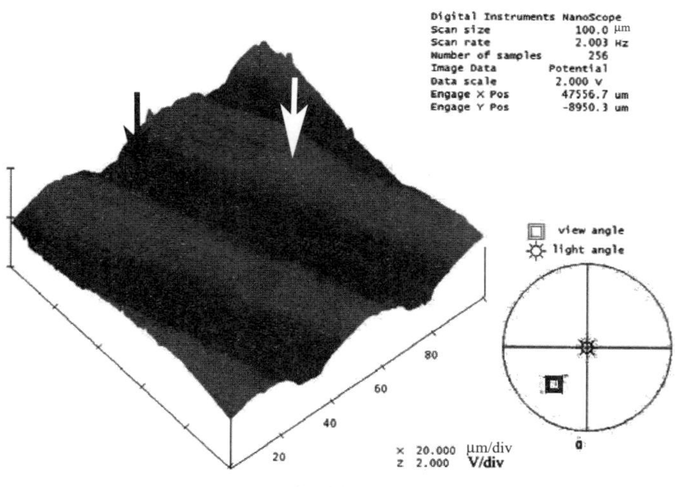


**Figure 12.** Residual surface potential of the gripper after application of 150V clamping voltage.

total potential difference is approximately 0.9 V. Over time, some of the trapped charge is neutralized through several possible mechanisms: recombination with airborne ions or ions generated by naturally occurring ionizing radiation, or surface and bulk ionic diffusion. Most frequently, dissociated water molecules are linked to oppositely charged species for the neutralization process. Based on potential measurements conducted at 5, 15, and 30 minutes after charge application (see Fig. 13), the estimated time constant for charge dissipation was 25 minutes. In the case of a Ni part, the voltage required

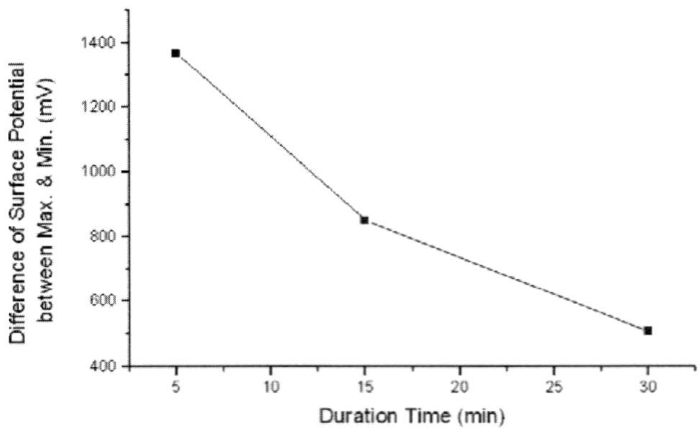


**Figure 13.** Charge decay time history.

to balance its own weight of 2.5 $\mu$N is 0.21V [from Eq. (3.6)]. It is therefore estimated that it will take up to 47 minutes to release the part under the action of its own weight, assuming that no significant van der Waals forces are present. Active release techniques are needed in order to successfully implement this technology. These include ultrasonic excitation (Bohringer et al., 1998), electromagnetic eddy current pulses (Feddema et al., 2002), thermal excitation (Arai and Fukuda, 1997), or rapid voltage reversal . While these efforts have produced improved release techniques, a reliable solution to this problem is yet to be found.

### 3.4 Magnetic Manipulation

Magnetic manipulation of micro-particles and objects has been explored for almost a century. In particular, observation of the motion of magnetic beads under a known magnetic force field has been used to determine viscoelastic properties of gels (Freundlich and Seifriz, 1923), as well as cytoplasm of biological cells (Crick and Hughes, 1950; Zaner and Valberg, 1989). More recently, Dietrich et al. (2004) demonstrated catheter steering using externally applied magnetic fields generated with permanent magnets. Naturally, magnetic manipulation of MEMS structures has also been explored by several groups. Judy and Muller (1997) used a magnetic force field to erect polycrystalline structures coated with a ferromagnetic layer and Ahn et al. (1996) developed a magnetic particle sorter. A novel application of the magnetic manipulation technique has been pioneered by Yesin et al. (2005), who developed a magnetically steered micro-robot assembled from individual ferromagnetic components. These were produced by a photolithographic process and subsequently assembled into a structure resembling a prolate spheroid (see Fig. 14.) When subjected to an external magnetic field, this structure is magnetized

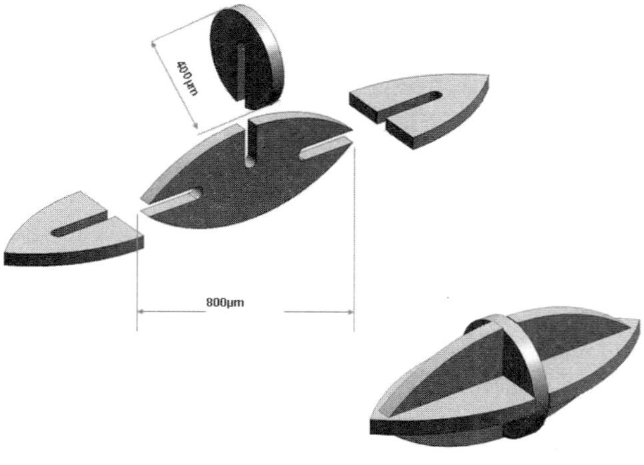


**Figure 14.** Ferromagnetic micro-robot assembled from electroplated components (Yesin et al., 2005). Courtesy of Prof. Nelson).

and develops magnetic force and moment according to

$$\mathbf{F}_m = \int_{V_m} \nabla(\mathbf{M} \cdot \mathbf{B}) dv \tag{3.11}$$

$$\mathbf{M}_m = \int_{V_m} \mathbf{M} \times \mathbf{B} dv, \tag{3.12}$$

respectively, where $V_m$ is the volume of the magnetized object, $\mathbf{B}$ is the magnetic flux density, and $\mathbf{M}$ is the magnetization of the object. The object is magnetized using an external magnetic field, $\mathbf{H}$, produced by several coils. For a linear isotopic material, the magnetization vector is proportional to the magnetic field strength

$$\mathbf{M} = \chi_m \mathbf{H}, \tag{3.13}$$

where $\chi_m$ is the material's magnetic susceptibility. The magnetic flux density then becomes

$$\mathbf{B} = \mu_0(1 + \chi_m)\mathbf{H} = \mu_0 \mu_r \mathbf{H}, \tag{3.14}$$

where $\mu_0$ is the magnetic permeability of the vacuum, and $\mu_r$ is the relative permeability of the object. Because of the prolate shape of the micro-part, the magnetization vector $\mathbf{M}$ tends to align itself with the major axis of the object (shape anisotopy effect (Aharoni and Shtrikman, 1957)). This effect aids in the development of torque according to Eq. (3.12), even though the electroplated material could be macroscopically isotropic. To independently control the force and torque, Yesin et al. (2005) used two pairs of co-axial coils. One of the pairs was driven by currents in the same direction, producing a uniform magnetic field in the space between the coils (Helmholtz coils), while the other pair was driven with opposing currents, producing uniform gradient of the magnetic field (Maxwell coils). Controlling the field and its gradient independently, it was possible to steer and propel the micro-robot through a two-dimensional maze filled with fluid.

## 4 Nano-Assembly Techniques

Nano-assembly of macromolecules and inorganic particles is an area of active research with great potential for application in bio- and nano-technology. Several approaches have been explored for the fabrication of organized molecules on a nanometer scale. The Langmuir-Blodget (LB) technique can be used for the transfer of organized monolayers of amphiphilic molecules from an air-water interface onto a solid substrate (Meyer et al., 1991), self-assembled monolayers (SAMs) of molecules with proper end-groups can be accomplished by chemisorption (Schoenfisch and Pemberton, 1998); and layer-by-layer deposition of films can be done electrostatically. In the last method, the Coulombic attraction between physisorped molecules and oppositely charged counter molecules in dilute solutions is utilized (Tien et al., 1997).

While these deposition methods form the basis of many industrially relevant processes, the only techniques available for pattern formation in the deposited films are photolithography (Sorribas et al., 2002), soft-lithography (PDMS stamping) (Whitesides et al., 2001), molecular imprinting (Cheng et al., 2001), and dip-pen nanolithography (DPN) (Wilson et al., 2001). The first two techniques utilize standard ultraviolet or e-beam lithography, resulting in relatively large patterns with minimum feature sizes of approximately 100 nm. The molecular imprinting technique temporarily incorporates larger molecules (template molecules) in the self-assembled monolayer. These are subsequently washed off, leaving behind an imprint of the template molecules. Although very promising, this technique is plagued by surface diffusion, resulting in the gradual destruction of the molecular imprints. Nano-assembly by means of dip-pen nanolithography (DPN) uses an atomic force microscope tip to pattern the molecules onto the surface in a desired size and shape ranging in width from 30-800 nm up to 100 $\mu$m in length (Wilson et al., 2001).

Recent progress in the assembly of nano-particles has resulted in a new method based on injected charges (Enikov and Palaria, 2004; Palaria and Enikov, 2006; Naujoks and Stemmer, 2005; Welle and Jacobs, 2005). This technique opens the possibility of scaling up the nano-assembly to practically viable manufacturing technique. The assembly process, also dubbed nanoxerography, has been demonstrated in liquid and gas phases (Naujoks and Stemmer, 2005; Welle and Jacobs, 2005). The technique was first suggested by Wright and Chetwynd (1998) in their review paper inspired by the work of Green et al. (1997), who demonstrated dielectrophoretic trapping of sub-micrometer latex spheres and bioparticles between a polynomial electrode array.

Molecular dielectrophoresis of avidin and DNA was also demonstrated around the same time (Bakewell et al., 1998). Later, Fudouzi et al. (2002) demonstrated the assembly of silica particles with a resolution of approximately 5 $\mu$m using electrified patterns created by a focused ion beam (FIB) writing on a calcium titanate substrate. Krinke et al. (2001) demonstrated 100-nm patterns of gold particles deposited from a gas phase onto a silicon dioxide substrate. Barry et al. (2003) developed PDMS stamps to be used as electrodes for electrifying patterns onto PMMA substrates. Recently, Naujoks and Stemmer (2004) demonstrated that a xerography-like process allows the direct fabrication of sub-micron spots of bio-molecules from an insulating oil phase (FC-77). Here, we review the underlying phenomena utilized in this assembly process.

### 4.1 Charge Writing for Nano-Assembly

The electrostatically-assisted nano-assembly protocol consists of two steps. During the first step, charged spots are deposited into a dielectric substrate. Subsequently, the substrate is exposed to a suspension of nano-particles, which assemble under the action of the non-uniform electrostatic field. Figure 15 illustrates the charging process. An

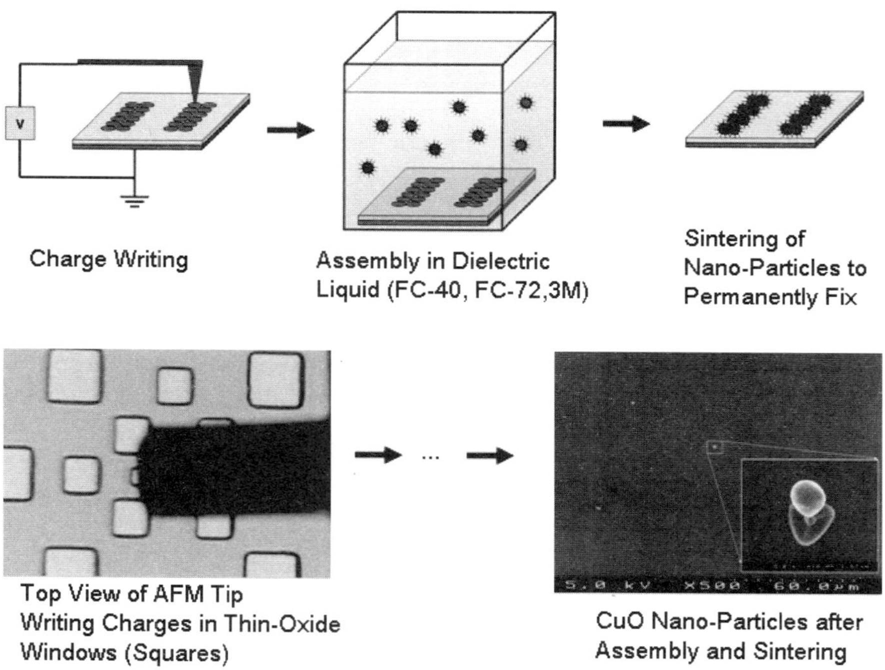


**Figure 15.** Assembly of nano-particles via charge writing with an AFM tip.

atomic force microscope (AFM) tip is used to inject a space-limited charge into the silicon dioxide. The silicon substrate is electrically connected to the grounded chuck of the AFM microscope. The tip is then biased with respect to the substrate at a fixed potential for a given amount of time, $\tau$. During this time, a charge is injected into the substrate through tunneling (Raju, 2003). Using the Fowler-Nordheim tunneling law, Palaria and Enikov (2006) developed a one-dimensional (1D) model for the charging process that can be used to estimate the strength of the field prior to nano-assembly. The model is based on Gauss' law, the charge conservation law, and the tunneling current law. The Gauss' law in 1D is given by

$$\epsilon_0 E_o - \epsilon_s E_s = -q n_t, \tag{4.1}$$

where $E_o$ is the electric field in the oxide (assumed constant), $E_s$ is the electric field at the silicon surface, $q$ is the elementary charge, and $n_t$ is the filled electron trap surface density. Assuming that no de-trapping occurs, the charge conservation law requires that the entire tunneling current is consumed for filling the traps, which leads to

$$\frac{dn_t}{dt} = (N_t - n_t)\frac{J\sigma}{q}, \tag{4.2}$$

where $N_t$ is the initial empty trap density and $\sigma$ is the trap cross section. The tunneling current is given by the Fowler-Nordheim model (Raju, 2003)

$$J = C_0 E_o^2 e^{-C_E/E_o}, \tag{4.3}$$

where $J$ is the current density, $E_o$ is the electric field in the oxide, and $C_0$ and $C_E$ are the Fowler-Nordheim constants given by

$$C_0 = \frac{q^2}{3.36\pi h \phi_B} \text{ and } C_E = \frac{8\pi\sqrt{0.84 m q \phi_B^3}}{3h}, \tag{4.4}$$

where $h$ is the Planck's constant, $\phi_B$ is the potential barrier height (see energy band-diagram in Fig. 16), and $q$ and $m$ are the charge and the electron rest mass, respectively. The electric field in the oxide, $E_o$, can be related to the applied potential to the AFM tip, $V$, and the surface potential of the Si by

$$V = \phi_s - E_o h_o, \tag{4.5}$$

where $\phi_s$ is the surface potential of silicon, $h_o$ is the thickness of the oxide layer. $E_s$ can be expressed in terms of $\phi_s$ using hole accumulation conditions at the Si surface (Muller and Kamins, 1986)

$$E_s = \sqrt{\frac{2qN_a}{\epsilon_s}}\sqrt{\frac{kT}{q}\left(e^{-q\phi_s/kT} - 1\right) + \phi_s}. \tag{4.6}$$

Substituting $E_s$ from Eq. (4.6) and and $E_o$ from Eq. (4.1) into Eq.(4.5) gives

$$\phi_s - \left[\left(\epsilon_s\sqrt{\frac{2qN_a}{\epsilon_s}}\sqrt{\frac{kT}{q}\left(e^{-q\phi_s/kT} - 1\right) + \phi_s} - qn_t\right)/\epsilon_0\right] h_o + V = 0, \tag{4.7}$$

which is an implicit equation for the silicon surface potential $\phi_s$. Upon solution, one can use Eq.(4.5) to obtain the electric field in the oxide. Subsequently Eqs.(4.3) and (4.2) can be used to obtain the time required for filling the traps, as shown in Figure 17.

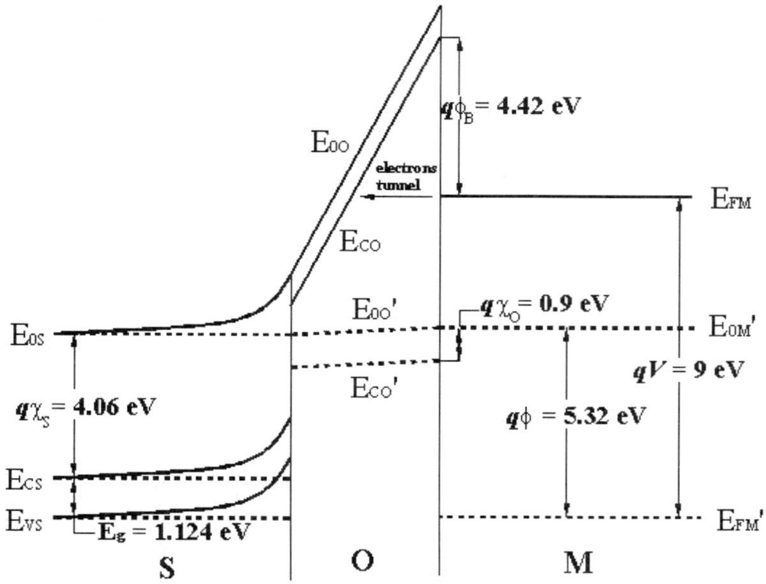


**Figure 16.** Energy band diagram of MOS structure representing the charge writing setup with 9 V applied to the metal electrode. $E_{0x}$ represents vacuum level, $E_{Cx}$ represents the lower edge of the conduction band, $E_{Vx}$ the upper edge of the valence band, and $E_{Fx}$ the Fermi level; $\chi$ is the electron affinity, $E_g = E_{Cx} - E_{Vx}$ the silicon band-gap, $\phi$ the metal work-function, V the applied potential bias, and $\phi_B$ the tunneling potential barrier. The sub-subscripts $x = S, O$, and $M$ denote semiconductor, oxide, and metal, respectively; $\phi_B$ is deduced as $\phi_B = V - \chi_O$. The dotted lines (primed notation) show the position of the bands before the application of the bias potential and the solid lines after biasing.

## 4.2 Driving Forces for Nano-Assembly

Gas phase and liquid phase assembly are the two main media used to transport the nano-particles to the assembly site. Assembly in the gas phase usually requires complex apparatus, including aerosol production using evaporation (Krinke et al., 2001) or electro-spraying (Welle and Jacobs, 2005), followed by charge neutralization to reduce the charge on the electro-sprayed particles, and an electrostatic steering chamber to guide the particles to the assembly sites (Welle and Jacobs, 2005). Alternatively, assembly in a liquid phase is much simpler and involves submersing the charged sample in a dielectric liquid containing the nano-particles. Only highly dielectric non-polar liquids are suitable for this technique, which limits the choice of dielectric fluids to mostly perfluoro-carbons, for example perfluorohexane ($C_6F_{14}$). Limited information is available on the interaction between this liquid and the nano-particles. Fudouzi et al. (2002) estimated the surface charge, $Q$, of the particles from measurements of their zeta potential, $\zeta$, which are related

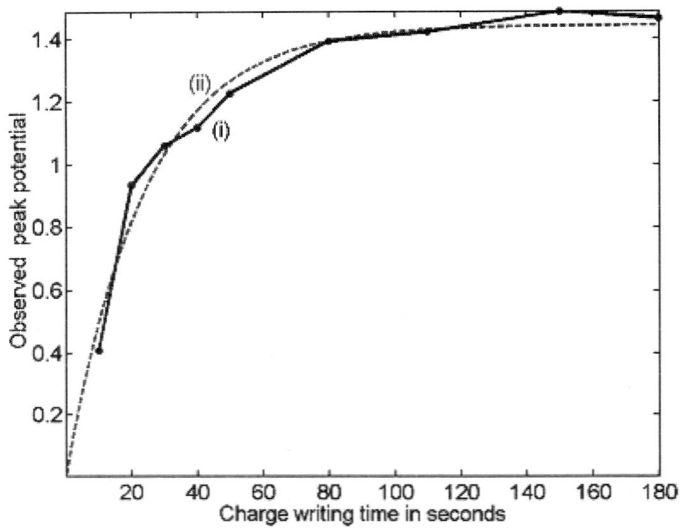


**Figure 17.** Plots of (i) the experimentally obtained peak potential and (ii) the scaled potential due to a simulated trapped charge extending over a radius of 150 nm vs. charge writing time (for comparison). The scaling factor for (ii) is 1.15.

through the following relation

$$Q = 4\pi R\nu\epsilon_l\epsilon_0\zeta \tag{4.8}$$

where $\nu$ is the viscosity of the liquid, $\epsilon_l$ is its dielectric constant, and $R$ is the radius of the nano-particle. The measured zeta potential was -25 mV, corresponding to -7×$10^{-18}$ C/particle of particle charge. If the particles do not possess a net charge and have different dielectric properties than the surrounding medium, a dielectrophoretic force, $F_{\text{dep}}$, created by a non-uniform external field can drive the assembly according to the well-known Clausius-Mossotti equation (Hughes, 2000)

$$F_{\text{dep}} = 2\pi R^3\epsilon_l K\nabla E^2, \qquad K = \frac{\epsilon_p - \epsilon_l}{\epsilon_p + 2\epsilon_l} \tag{4.9}$$

where $K$ is the Clausius-Mossotti factor and $\epsilon_p$ is the dielectric constant of the particle. Under the combined action of these forces, the dispersed nano-particles assemble onto the charged sites. This is illustrated in the lower half of Figure 15

## 5 Summary

This chapter represents a brief overview of the main developments in micro-assembly techniques for micro-systems as presented in the actual lectures during the CISM short course on Micro-Mechanical Systems Design (June 28 – July 2, 2004). Due to its future importance, an additional section on nano-assembly has been included in the printed version of these lecture notes. The material in this chapter does not constitute an exhaustive review of all micro- and nano-assembly techniques, but merely attempts to capture the experience of the author in the use of several such techniques for micro- and nano-assembly.

## Acknowledgement

This material is based upon the work supported by the National Science Foundation under Grants No. DMI-0134585 and DMI-0303868. Any opinions, findings, and conclusions or recommendations expressed in this material are those of the author(s) and do not necessarily reflect the views of the National Science Foundation.

## Bibliography

Aharoni, A. and Shtrikman, S. (1957) Magnetization curve of the infinite cylinder. *Physical Review*, 109(5), pp. 1522–1528.

Ahn, C., Allen, M., Trimer, W., Jun, Y.N., and Erramilli, S. (1996) A fully integrated micromachined magnetic particle manipulator and separator. *Journal of Microelectromechanical Systems*, 5(3), pp. 151–158.

Arai, F., Andou, D., and Fukuda, T. (1996) Adhesion forces reduction for micro manipulation based on micro physics. In *Proceedings of the IEEE Workshop on Micro Electro Mechanical Systems*. pp. 354–359.

Arai, F. and Fukuda, T. (1997) A new pick up and release method by heating for micromanipulation. In *Tenth Annual International IEEE Workshop on Micro Electro Mechanical Systems*. pp. 383–388.

Bakewell, D., Hughes, M., Milner, J., and Morga, H. (1998) Dielectrophoretic manipulation of avidin and dna. In *Proc. $20^t h$ Annual International Conference of the IEEE Engineering in medicine and Biology Society Advanced Materials*, vol. 20. IEEE, pp. 1079–1082.

Bao, M.H. (2000) *Micro Mechanical Transducers: Pressure Sensors, Accelerometers and Gyroscopes:Handbook of Sensors and Actuators, v. 8, Ed. S, Middelhoek*. Handbook of Sensors and Actuators. Elsevier.

Barry, C.R., Steward, M.G., Lwin, N.Z., and Jacobs, H.O. (2003) Printing nanoparticles from the liguid and gas phases using nanoxerography. *Nanotechnology*, 14, pp. 1057–1063.

Bellouard, Y., Clavel, R., Gotthardt, R., Bidaux, J., and Sidler, T. (1998) A new concept of monolithic shape memory alloy micro-devices used in micro-robotics. In *Actuator'98- 6th International Conference on New Actuators - Bremen, Germany, June 17-19, 1998*. IEEE, pp. 499–502.

Bohringer, K.F., Goldberg, K.Y., Cohn, M., Howe, R., and Pisano, A. (1998) Parallel microassembly with electrostatic force fields. In *Proceedings of the IEEE International Conference on Robotics and Automation (ICRA-98)*. IEEE, pp. 1204–1211.

Bowling, R.A. (1988) A theoretical review of particle adhesion. In *Particles on Surfaces 1: Detection, Adhesion and Removal*. Plenum Press, pp. 129–155.

Breguet, J.M., Henein, S., Mericio, R., and Clavel, R. (1997) Monolithic piezoceramic flexible structures for micromanipulation. In *Proc of the 9th International Precision Engineering Seminar*. pp. 397–400.

Carrozza, M., Dario, P., Menciassi, A., and Fenu, A. (1998) Manipulating biological and mechanical micro-objects using liga-microfabricated end-effectors. In *Proceedings of the 1998 IEEE International Conference on Robotics C Automation Leuven, Belgium - May 1998*. IEEE, pp. 1811–1816.

Cheng, Z., Wang, E., and Yang, X. (2001) Capacitive detection of glucose using molecularly imprinted polymers. *Biosensors & Bioelectronics*, 16, pp. 179–185.

Cohn, M.B. and Howe, R.T. (2000) Wafer-to-wafer transfer of microstructures using breakaway tethers. *US Patent 6142358*.

Crick, F. and Hughes, A. (1950) The physical properties of cytoplasm a study by means of the magnetic particle method. 1. experimental. *Experimental cell research*, 1(1), pp. 37–80.

Dietrich, T., Kleen, M., Killmann, R., Wiesinger, B., Wiskirchen, J., Tepe, G., Claussen, C., and Duda, S. (2004) Evaluation of magnetic navigation in an in vitro model of uterine artery embolization. *Journal of vascular and interventional radiology*, 15(12), pp. 1457–1462.

Enikov, E.T. and Lazarov, K.V. (2003) An optically transparent gripper for micro-assembly, journal of micromechatronics. *Journal of Micromechatronics*, 2(2), pp. 121–140.

Enikov, E.T. and Lazarov, K.V. (2005) Micro-mechanical switch array for meso-scale actuation. *Sensors and Actuators A*, 121(1), pp. 282–293.

Enikov, E.T., Minkov, L.L., and Clark, S. (2005) Micro-assembly experiments with transparent electrostatic gripper under optical and vision-based control. *IEEE Transactions on Industrial Electronics*, 52(4), pp. 1005–1012.

Enikov, E.T. and Palaria, A. (2004) Charge writing in silicon-silicon dioxide for nano-assembly. *Nanotechnology*, 15(9), pp. 1211–1216.

Feddema, J., Ogden, A., Warne, L., Johnson, W., and Armour, D. (2002) Electrostatic/electromagnetic gripper for micro-assembly. *Sandia technical report SAND2002-0490C*.

Freundlich, H. and Seifriz, W. (1923) On the elasticity of soles and gels. *Zeitschrift fur Physikalische Chemie Stochiometrie und Verwandtschaftslehre*, 104(3-4), pp. 233–261.

Fudouzi, H., Kobayashi, M., and Shinya, N. (2002) Site-controlled deposition of micro-sized particles using an electrostatic assembly. *Advanced Materials*, 14, pp. 1649–1652.

Gengenbach, U. and Boole, J. (2000) Electrostatic feeder for contactless transport of miniature and microparts. In *Proceedings of SPIE Microrobotics and Microassembly II, Eds. Bradley J. Nelson, Jean-Marc Breguet*, vol. 4194. SPIE, pp. 75–81.

Green, N., Morgan, H., and Milner, J. (1997) Manipulation and trapping of sub-micron bioparticles using dielectrophoresis. *Journal of Biochemical and Biophysical Methods*, 35, pp. 89–102.

Haliyo, D., Rollot, Y., and Regnier, S. (2001) Dynamical strategies for the micro-manipulation by adhesion. In *Proceedings of SPIE ,vol. 4568 Microrobotics and Microassembly III, Edited by B. J Nelson and J.M. Breguet*. pp. 261–269.

Hara, S., Nakada, H., Sawada, R., and Isomura., Y. (1993) High precision bonding of semiconductor laser diodes. *Int. J. Japan Soc. Prec. Eng*, 27(1), pp. 49–53.

Hesselbach, J., Buttgenbach, S., Wrege, J., Butefisch, S., and Graf, C. (2003) Centering electrostatic microgripper and magazines for microassembly tasks. In *Proceedings of SPIE Microrobotics and Microassembly III, Eds. Bradley J. Nelson, Jean-Marc Breguet*, vol. 4568. SPIE, pp. 270–277.

Hesselbach, J., Pittschellis, R., Hornbogen, E., and Mertmann, M. (1997) Shape memory alloys for use in miniature grippers. In *Proc SMST, Int. Conf. On Shape Memory and Superelastic Technologies*. pp. 251–256.

Hughes, M. (2000) Ac electrokinetics applications for nanotechnology. *Nanotechnology*, 11, pp. 124–132.

Israelachvili, J. (1991) *Intermolecular and Surface Forces*. Academic Press, London.

Jacobs, H.O., Knapp, H.F., and Stemmer, A. (1999) Practical aspects of kelvin probe force microscopy. *Review of Scientific Instruments*, 70(3), pp. 1756–1760.

Jacobs, H.O., Tao, A.R., Schwartz, A., Gracias, D.H., and Whitesides, G.M. (2002) Fabrication of a cylindrical display by patterned assembly. *Science*, 296.

Judy, J.W. and Muller, R. (1997) Magnetically actuated, addressable microstructures. *Journal of Microelectromechanical Systems*, 6(3), pp. 249–256.

Kamura, Y. and Yanagimachi, R. (1995) Intracytoplasmic sperm injection in the mouse. *Biology of Reproduction*, 52, pp. 709–720.

Keller, C. and Howe, R. (1995) Nickel-filled hexsil thermally actuated tweezers. In *Transducers 95:Eighth International Conference on Solid-State Sensors and Actuators*. pp. 376–379.

Kim, C., Pisano, A., Muller, R., and Lim, M. (1990) Polysilicon microgripper. In *Proc. IEEE Solid-State Sensor and Actuator Workshop (Hilton Head Island, SC, USA*. IEEE, pp. 48–51.

Kohl, M., Krevet, B., and Just, E. (2002) Sma microgripper system. *Sensors and Actuators A*, 97-98, pp. 646–652.

Krinke, T.J., Fissan, H., Deppert, K., Magnusson, M.H., and Samuelson, L. (2001) Positioning of nanometer-sized particles on flat surfaces by direct deposition from the gas phase. *Appl. Phys. Lett.*, 78, pp. 3708–3710.

Lowell, J. and Rose-Innes, A.C. (1980) Contact electrification. *Advances in Physics*, 29(6), pp. 947–1023.

Meyer, E., Overney, E.H.R., Heinzelmann, H., Frommer, J., Guntherodt, H., Wagner, T., Schiert, H., and Roth, S. (1991) Molecular-resolution images of langmuir?blodgett films using atomic force microscopy. *Nature*, 349, pp. 398–400.

Mikkelson, A., Engeistada, R., Lovell, E., Aschke, L., Ruggeberg, F., and Sobel, F. (2004) Effect of electrostatic chucking on euvl mask flatness. In *Proceedings of SPIE Vol. 5504, 20th European Conference on Mask Technology for Integrated Circuits and Microcomponents, Ed. Uwe F. W. Behringer*, vol. 5504. SPIE, pp. 120–128.

Miyazaki, H. and Zato, T. (1996) Pick and place shape forming of three-dimensional micro structures from fine particles. In *Proceedings of the IEEE International Conference on Robotics and Automation (ICRA)*. pp. 2535–2540.

Muller, R.S. and Kamins, T.I. (1986) *Device Electronics for Integrated Circuits.* John Wiley and Sons, New York.

Naujoks, N. and Stemmer, A. (2004) Using local surface charges for the fabrication of protein patterns. *Colloids and Surfaces A: Physicochem. Eng. Aspects*, 249, pp. 69–72.

Naujoks, N. and Stemmer, A. (2005) Micro- and nanoxerography in liquids controlling pattern definition. *Microelectronic Engineering*, 78-79, pp. 331–337.

Nicollian, E. and Brews, J. (2003) *MOS (Metal Oxide Semiconductor) Physics and Technology.* Wiley-Interscience, New Jersey.

Oh, H.S. (1989) Elekrtosatische greifer für die mikromontage. *Fortschritt-Berichte VDI, Reihe 8, Mesteuerungs- und Regelungstechnik*, 8(702).

Palaria, A. and Enikov, E.T. (2006) Experimental analysis of the stability of electrostatic bits for assisted nano-assembly. *Journal of Electrostatics*, 64, pp. 1–9.

Raju, G. (2003) *Dielectrics in Electric Fields.* Dekker Inc., New York.

Schoenfisch, M. and Pemberton, J. (1998) Air stability of alkanethiol self-assembled monolayers on silver and gold surfaces. *Journal of the American Chemical Society*, 120, pp. 4502–4513.

Singh, A., Horsley, D.A., Cohn, M.B., Pisano, A.P., and Howe, R.T. (1999) Batch transfer of microstructures using flip-chip solder bonding. *Journal of Microelectromechanical Systems*, 8(1), pp. 27–33.

Sorribas, H., Padeste, C., and Tiefenauer, L. (2002) Photolithographic generation of protein micropatterns for neuron culture applications. *Biomaterials*, 23(3), pp. 893–900.

Tien, J., Terfort, A., and Whitesides, G. (1997) Microfabrication through electrostatic self-assembly. *Langmuir*, 13, pp. 5349–5355.

Vikram, B., Nelson, B., Yang, G., and Enikov, E.T. (2000) Microassembly of hybrid magnetic mems. *Journal of Micromechatronics.* Accepted for publication June, 2000.

Welle, A.M. and Jacobs, H. (2005) Printing of organic and inorganic nanomaterials using electrospray ionization and coulomb-force-directed assembly. *Applied Physics Letters*, 87, p. 263119.

Whitesides, G.M., Ostuni, E., Takayama, S., Jiang, X., and Ingber, D. (2001) Soft lithography in biology and biochemistry. *Annual Reviews in Biomedical Engineering*, 3, pp. 335–373.

Wilson, D., Martin, R., Hong, S., Cronin-Golomb, M., Mirkin, C., and Kaplan, D. (2001) Surface organization and nanopatterning of collagen by dip-pen nanolithography. *Biophysics*, 28, pp. 13660–13664.

Wright, W. and Chetwynd, D. (1998) Can charge writing aid nanotechnological manipulation? *Nanotechnology*, 9, pp. 133–142.

Yesin, K., Exner, P., Vollmers, K., and Nelson, B. (2005) Design and control of in-vivo magnetic microrobots. In *Medical Image Computing and Computer-Assisted Intervention MICCAI2005, Pt.1/Lecture Notes in Computer Science*, vol. 3749. Springer, pp. 819–826.

Zaner, K. and Valberg, P. (1989) Viscoelasticity of f-actin measured with magnetic microparticles. *Journal of cell biology*, 109(5), pp. 2233–2243.

# Microfluidics: Device Science and Technology

Lap Man Lee, Luthur Siu Lun Cheung and Yitshak Zohar

Department of Aerospace & Mechanical Engineering, University of Arizona, Arizona, USA

**Abstract.** This paper presents the basic concepts related to the newly emerging field of microfluidics. Following a brief introduction of the general conservation and particular laws, three size effects are introduced. The velocity slip boundary condition for gas flows as well as the electrokinetic and polar-mechanics effects for liquid flow in microdomains are introduced. Pressure-driven gas flows and electrokinetically-driven liquid flows in microchannels are analyzed in details. Finally, several flow diagnostic techniques and fabrication of microfluidic systems are described.

## 1 Introduction

With the start of the new millennium, the subject of microfluidics has become an integral part of several research areas, most notably in bio applications, becoming the foundation of a new field popularly called Bio-MEMS. A fundamental requirement for the commercial success of any microfabrication technology, e.g. microfluidics, is an application with a very large demand. Such an application is an essential technology driver that provides sufficient economic pull for the adequate recovery of infrastructure investments to sustain continued research into new and improved devices at very low unit cost. A new application for MEMS in general and microfluidics in particular that not only satisfy this requirement but also promises enormous potential for growth is biotechnology. Such a broad base application may prove to be a decisive technology driver for microfluidics.
Several titles have already appeared in print addressing theoretical/numerical (Karniadakis and Beskok 2002) or devices/technology (Nguyen and Wereley 2002) aspects of microfluidics. As in any emerging field, what exactly constitute a microfluidic system is not well-defined. Certainly, the term 'micro' suggests a length scale on the order of a micrometer, while 'fluidics' indicates a system in which gas, liquid or a mixture of both is involved either in a primary or secondary role. Clearly, if fluidic microsystems are to be considered different from macrosystems, it means that properties of microscale flows cannot be deduced from their macro counterparts by simple scaling. Due to the size effect, some fluid phenomena negligible at macro domains become dominant in micro domains and vise versa. The field of microfluidics was initially thought to be simple as flow phenomena such as turbulence, flow separation or hydrodynamic instabilities were to be suppressed due to the close proximity of the solid boundaries. However, it has quickly become apparent that flows in microsystems can be very complex, despite the stabilizing effect of the walls, due to the dominant role of the fluid/solid interaction on the flow development. Hence, theoretical models and numerical simulations of fluidic microsystems have to account for such effects in order to facilitate adequate comparisons with experimental results when available.

The field of microfluidics embraces not only the science of fluid flows in microscale but also the microfabrication technologies required to realize such microsystems as well as the diagnostic techniques needed to measure various flow properties. Microfluidic systems stand to benefit from numerous advantages including: higher spatial resolution, smaller time constant, and batch fabrication to ultimately lower the cost of commercial devices and systems. An attractive feature of microsystems is the potential for high degree of integration of various components on a single chip. The possibility of integrating sensors and actuators with electronic circuits holds the promise of manufacturing microfluidic systems with sensing, decision-making and action capabilities.

Fluid flows, whether macro or micro, have an enormous scope of applications and are classified in different categories based on distinct characteristic features. Internal flows are bounded by solid boundaries while external flows typically engulf a solid body and theoretically extend to infinity. Free flows such as jets or wakes evolve without an interaction with a solid boundary. Multi-phase flows, as opposed to single-phase flow, involve the flow of matter in at least two phases. Prominent among them are two phase flows of liquid-vapor mixtures with the complexity of interfacial phenomena, and suspension flows of solid particles immersed in liquid flows. Experience indicates that it is highly constructive to select simple case-studies in order to demonstrate complex effects. Therefore, pressure-driven gas flows and electrokinetically-driven liquid flows in microchannels have been selected illustrate the size effects in microfluidic systems. A brief description of several diagnostic techniques as well as the fabrication of a few fluidic microsystems is also included.

# 2 Fundamentals of Microscale Fluid Mechanics

## 2.1 The continuum hypothesis

Even though fluids seem to be continuum according to our observation in general, they are however composed of many finite-size molecules with finite distance between them. These molecules are in constant random motion and collisions. Hence, fluids are no longer continuum but composed of individual particles when we scale down our observation volume to be sufficiently small. But, how large should the observation volume be? If the observation volume is large enough to contain a sufficient number of molecules of the fluid to give an average value for any property that is continuous in space and independent of the number of molecules, the fluid can be treated as continuous such that the density and velocity are continuous and differentiable in space and time.

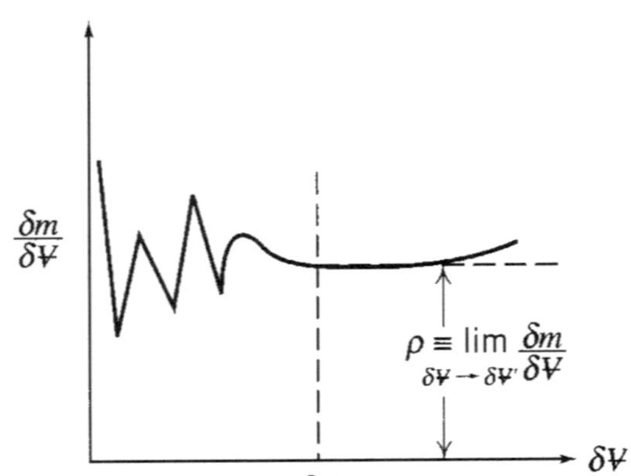


**Figure 2.1.1.** The average density as a function of the sampling volume.

## 2.2 Mean free path

The molecular scale in gases most relevant to transport of momentum and energy is the mean free path which is defined as the average distance traveled by the molecules between successive collisions; this is different from the average distance between molecules. Base on kinetic theory of gases, the mean free path $\lambda$ can be calculated as a function of the gas properties and conditions as follows:

$$\lambda = \frac{\mu}{P}\left(\frac{\pi RT}{2}\right)^{1/2} \tag{2-1}$$

The non-dimensional form of mean free path, Knudsen number $Kn$, is regarded as the parameter for determining the validity of the continuum assumption defined as:

$$Kn \equiv \frac{\lambda}{L} \tag{2-2}$$

If $Kn$ approaches zero, the system can be classified as continuum mechanics problem because the characteristic length scale is much larger than the mean free path. Flows are classified among different regimes based on $Kn$ as follows (Schaaf and Chambre 1961):

**Table 2.1.1.** Classification of the different gas flow regimes.

| | |
|---|---|
| $Kn < 10^{-3}$ | Continuum Flow |
| $10^{-3} < Kn < 10^{-1}$ | Slip Flow |
| $10^{-1} < Kn < 10^{1}$ | Transition Flow |
| $10^{1} < Kn$ | Free Molecular Flow |

## 2.3 Thermodynamic concepts (Equilibrium)

The most convenient framework within which fluid mechanics problems can be studied is the system, which is a quantity of matter not necessarily constant, contained within a boundary. The boundary can be physical, partly physical and partly imaginary, or wholly imaginary. The physical laws to be discussed are always stated in terms of a system. A control volume is any specific region in space across the boundaries of which mass, momentum and energy may flow, and within which mass, momentum and energy storage may take place, and on which external forces may act. The complete definition of a system or a control volume must include, implicitly at least, the definition of a coordinate system, since the system may be moving or stationary. The characteristic of a system of interest is its state, which is a condition of the system described by its properties. A property of a system can be defined as any quantity that depends on the state of the system and is independent of the path, i.e. previous history, by which the system arrived at the given state. If all the properties of a system remain unchanged, the system is said to be in an equilibrium state.

A change in one or more properties of a system necessarily means a change in the state of the system has occurred. The path of the succession of states through which the system passes is called the process. When a system in a given initial state goes through a number of different changes of state or processes and finally returns to its initial state, the system has undergone a cycle. The properties describe the state of a system only when it is in equilibrium. If no heat transfer takes place between any two systems when they are placed in contact with each other, they are said to be in thermal equilibrium. Any two systems are said to have the same temperature if they are in thermal equilibrium with each other. Two systems that are not in thermal equilibrium have different temperature, and heat transfer may take place from one system to the other. Therefore, temperature is a property, which measures the thermal level of a system.

### 2.4 General Laws

A physical law is called a general law if its application is independent of the medium under consideration. Otherwise, it is called a particular law. The three general laws, upon which all the analyses concerning fluid mechanics depend, are:
- Conservation of mass
- Conservation of momentum
- Conservation of energy

**Conservation of mass:** The law of conservation of mass simply states that, in the absence of any mass-energy conversion, the mass of the system remains constant. Thus, in the absence of a source or a sink, the rate of change of mass in a control volume is equal to the mass flux through the control surface.

$$\frac{\partial}{\partial t}\int_{CV}\rho d\upsilon + \int_{CS}\rho(\mathbf{U}\cdot\mathbf{n})dA = 0 \tag{2-3}$$

$$\frac{D\rho}{Dt} + \rho(\nabla\cdot\mathbf{U}) = 0 \tag{2-4}$$

For the time independent problem, steady-state flow,

$$\nabla\cdot(\rho\mathbf{U}) = 0 \tag{2-5}$$

For the time independent problem with incompressible substance,

$$\nabla\cdot\mathbf{U} = 0 \tag{2-6}$$

**Conservation of momentum:** Newton's second law of motion states that the sum of the external forces, $\Sigma F$, acting on a system in an inertial coordinate system is equal to the time rate of change of the total linear momentum of the system.

$$\frac{\partial}{\partial t}\int_{CV}\rho\mathbf{U}d\upsilon + \int_{CS}\rho\mathbf{U}(\mathbf{U}\cdot\mathbf{n})dA = \sum\mathbf{F} \tag{2-7}$$

$$\rho\frac{D\mathbf{U}}{Dt} = -\nabla P + \rho\mathbf{f} + \mu\nabla^2\mathbf{U} \tag{2-8}$$

Equation 2-8 is the Navier-Stokes equation.

**Conservation of energy:** The first law of thermodynamics, which is a particular statement of conservation of energy, states that the rate of change in the total energy of a system, $E$, undergoing a process is equal to the difference between the rate of heat transfer to the system, $Q$, and the rate of work done by the system, $W$.

$$\frac{\partial}{\partial t}\int_{CV} \rho e d\upsilon + \int_{CS} \rho e(\mathbf{U}\cdot\mathbf{n})dA = \frac{dE}{dt} = \dot{Q} - \dot{W} \tag{2-9}$$

$$\rho c_p \frac{DT}{Dt} = k\nabla^2 T + \Phi \tag{2-10}$$

Where dissipation due to friction, $\Phi$, is given by,

$$\Phi = 2\left[\left(\frac{\partial u}{\partial x}\right)^2 + \left(\frac{\partial v}{\partial y}\right)^2 + \left(\frac{\partial w}{\partial z}\right)^2\right] + \left(\frac{\partial v}{\partial x} + \frac{\partial u}{\partial y}\right)^2 + \left(\frac{\partial w}{\partial y} + \frac{\partial v}{\partial z}\right)^2 + \left(\frac{\partial u}{\partial z} + \frac{\partial w}{\partial x}\right)^2 \tag{2-11}$$

## 2.5 Particular Laws

**Heat convection:** Newton's law of cooling states that the heat flux from a solid surface to the ambient fluid by convection, $q''$, is proportional to the temperature difference between the solid surface temperature, $T_w$, and the fluid free-stream temperature, $T_\infty$, as follows,

$$q'' = h(T_w - T_\infty) \tag{2-12}$$

Where $h$ is the heat transfer coefficient.

**Equation of state:** A system in thermodynamic equilibrium is incapable of spontaneous change as it is in complete balance with the surroundings. The state of a pure substance at equilibrium is defined by two independent properties. All other properties are then uniquely determined in terms of these two properties. They may be found either by experiment or by the used of suitable equations of state. For example, it has been established that the behaviour of gases at low density is closely given by the ideal-gas equation of state:

$$P = \rho RT \tag{2-13}$$

where R is the specific gas constant. At very low-density, all gases and vapours approach ideal-gas behaviour. However, the behaviour may deviate substantially from that at higher densities.

**Size effects:** Length scale is a fundamental quantity that dictates the type of forces or mechanisms governing physical phenomena. Miniaturization of systems, therefore, involves not new physical laws or forces but rather different physical behaviour due to the difference in relative contribution of various forces or mechanisms. The large surface-to-volume ratio in micro devices accentuates the role of surface effects. Furthermore, the intrinsic length scale in gases, the mean free path, is much larger than the average distance between liquid molecules. Consequently, in microscale convection heat transfer, the size effects observed in single-phase gas or liquid flow and two-phase vapour/liquid flow are radically different from each other.

**Non-continuum mechanics:** The flow of a gas in the continuum regime can be thought of as the combination of a random molecular motion superposed on an ordered directional flow. Near a solid boundary, there are two possible ways in which a molecule may reflect from a stationary surface as illustrated in Figure 2.5.1. If the solid surface is perfectly smooth, the reflection is specular; the angle of incidence is equal to the angle of reflection with no change in the velocity component tangential to the surface. In a real case, however, the surface is not smooth but rather rough, and the reflection is diffuse; the molecules may move away from the wall at any angle. Arkilic et al. (1997) investigated that, for the diffuse reflection, on the average, the molecules essentially lose their tangential velocity component.

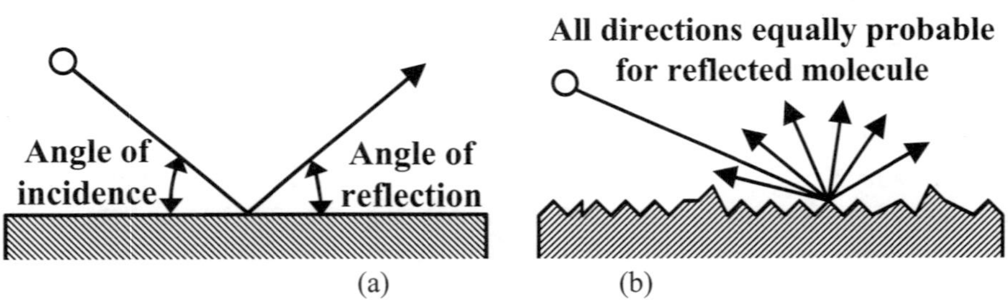


**Figure 2.5.1.** Schematic illustration of (a) specular, and (b) diffuse reflection.

If the mean free path becomes very large, intermolecular collisions become infrequent. Molecules arriving at a surface from a free stream may not have had a chance to be slowed down by collisions with other molecules; rather they arrive at the surface with free stream velocity, little affected by the presence of the solid surface. On the other hand, the molecules reflected diffusely from the wall have no net tangential velocity. Those reflected specularly have the same tangential velocity as the incident molecules. If $\sigma_T$ denotes the fraction of diffusely reflected molecules, Beskok and Karniadakis (1994) modeled the average velocity at the surface through Maxwell's velocity-slip condition:

$$U_s - U_w = \frac{2-\sigma_U}{\sigma_U} \lambda \left.\frac{\partial U}{\partial n}\right|_w \tag{2-14}$$

Similarly, the gas temperature at the solid surface is given by Smoluchowski's temperature-jump condition:

$$T_j - T_w = \frac{2-\sigma_T}{\sigma_T} \frac{2c_p}{c_p + c_v} \frac{k}{\mu c_p} \lambda \left.\frac{\partial T}{\partial n}\right|_w \tag{2-15}$$

$U_w$ and $T_w$ are the solid-surface velocity and temperature, while $U_s$ and $T_j$ are the gas-flow velocity and temperature at the boundary, respectively; n is the direction normal to the surface, while $c_p$ and $c_v$ are the specific heats. $\sigma_U$ and $\sigma_T$ are the tangential momentum and energy accommodation coefficients, which model the momentum and energy exchange of the gas molecules impinging on the solid boundary. They can vary from zero (specular accommodation) to one (complete, or diffuse accommodation). When gas molecules impinge on a surface, in general, the molecules exchange three properties with the surface: energy, normal momentum and tangential momentum. The extent to which impinging molecules reach equilibrium with the surface is represented by the accommodation coefficients.

The accommodation coefficients are empirically determined, and depend on both the gas molecules and the solid surface. Among the important parameters determining the reflection characteristics of a solid-gas interface is the surface roughness. If the surface is smooth on a molecular scale of the impinging gas, the reflection will be perfectly specular with no streamwise momentum transfer to the solid surface; this implies that the incident is equal to the reflected streamwise velocity and $\sigma_U$=0. If the surface is rough on a molecular scale, all the streamwise momentum will be transferred to the surface. This diffuse reflection is characterized by complete transfer of streamwise momentum with $\sigma_U$=1. The accommodation coefficients depend not only on the geometry of the surface but also on the molecular species adsorbed on the surface. For example, it was reported by Lord (1974) that partially removing the adsorbed gas layer in polished steel, with surface roughness of 0.1 μm, the tangential accommodation coefficient for helium could be reduced from near unity to about 0.8. However, by partially removing the adsorbed gasses, no similar reduction could be obtained for steel that was roughened. Although the exact nature of the accommodation coefficients is still an active research problem, almost all evidence indicates that for most gas-solid interactions the coefficients are approximately 1.0.

## 2.6 Electrokinetics

Most solid surfaces are likely to carry electrostatic charge, i.e. an electric surface potential, due to broken bonds and surface charge traps. When a liquid containing small amount of ions is forced through a microchannel under hydrostatic pressure, the solid-surface charge will attract the counter-ions in the liquid to establish an electric field. The arrangement of the electrostatic charges on the solid surface and the balancing charges in the liquid is called the Electric Double layer, EDL, as illustrated in Figure 2.6.1. Counter-ions are strongly attracted to the surface forming a compact layer, called the Stern layer, of immobile counter ions at the solid/liquid interface due to the surface electric potential. Outside this layer, the ions are affected less by the electric field and are mobile. The distribution of the counter-ions away from the interface decays exponentially within the diffuse double layer, called the Gouy-Chapman layer, with a characteristic length inversely proportional to the square root of the ion concentration in the liquid. The EDL thickness ranges from a few up to several hundreds of nanometers, depending on the electric potential of the solid surface, the bulk ionic concentration and other properties of the liquid. Consequently, EDL effects can be neglected in macrochannel flow. In microchannels, however, Yang (1998) reported the EDL thickness is often comparable to the characteristic size of the channel, and its effect on the fluid flow and heat transfer may not be negligible.

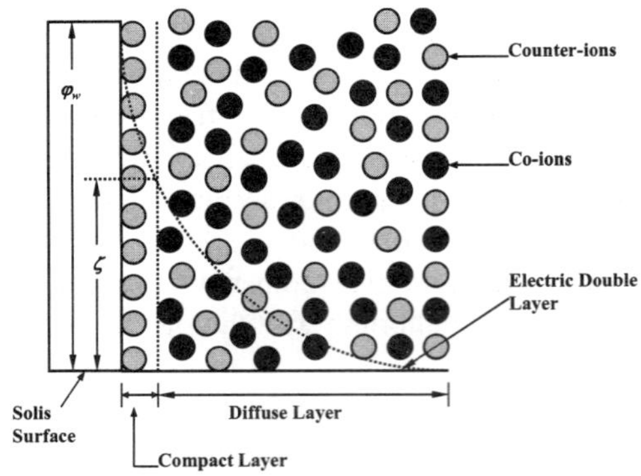


**Figure 2.6.1.** A schematic of the electric double layer (EDL) structure at the vicinity of the solid/liquid interface.

Consider a liquid between two parallel plates, separated by a distance $H$, containing positive and negative ions in contact with a planar positively charged surface. The surface bears a uniform electrostatic potential $\psi_0$, which decreases with the distance from the surface. Mala (1997) reported that the electrostatic potential, $\psi$, at any point near the surface is governed by the Debye-Huckle linear approximation:

$$\frac{d^2\psi}{dy^2} = \frac{\psi}{\lambda_D^2} \tag{2-16}$$

The characteristic thickness of the EDL is the Debye length, defined as

$$\lambda_D = \left( \frac{\varepsilon\varepsilon_0 k_b T}{2n_0 z^2 e^2} \right)^{1/2} \tag{2-17}$$

where $\varepsilon$ is the dielectric constant of the medium, $\varepsilon_0$ is the permittivity of vacuum, and $k_b$ is the Boltzmann constant. $z$ is the valence of negative and positive ions, $e$ is the electron charge, and $n_0$ is the ionic concentration. If the electrical potential of the channel surface is small and the separation distance between the two plates is larger than the thickness of the EDL ($H/2\lambda_D>1$) so that the EDLs near the two plates do not overlap, the appropriate boundary conditions are: $\psi=0$ at the mid-point ($y=0$) and $\psi=\psi_w$ on both walls ($y=\pm H/2$). The solution is then:

$$\psi = \varsigma \frac{|\sinh(y/\lambda_D)|}{\sinh(H/2\lambda_D)} \tag{2-18}$$

where $\varsigma$ is the zeta electric potential at the boundary between the diffuse double layer and the compact layer. The potential in the diffuse double layer results in an electric body force acting on fluid particles along with the pressure and viscous forces. This force could be significant in microchannel liquid flow, and should be included in the momentum equation. At the thin EDL limit $De>>1$ (with zero pressure-gradient), the velocity profile of an electroosmotic flow in a channel or a tube is approximately uniform (plug flow) given by:

$$u(y) = -\frac{\varepsilon\varepsilon_0 E\zeta}{\mu} \tag{2-19}$$

where $E$ is an externally applied electric field, while the Debye number is defined as:

$$De \equiv \frac{L}{\lambda_D} \tag{2-20}$$

## 2.7 Polar mechanics

In addition to the usual concepts of classical nonpolar fluid mechanics, there are two main physical concepts associated with polar mechanics: couple stresses and internal spin (Stokes, 1984). In nonpolar mechanics, the mechanical action of one part of a body on another is assumed to be equivalent to a force distribution only. However, in polar mechanics, the mechanical action is assumed to be equivalent to both a force and a moment distribution as illustrated in Figure 2.7.1. Thus, the state of stress at a point in nonpolar mechanics is defined by a symmetric second order tensor, which has six independent components. On the other hand, in polar mechanics, the state of stress is determined by a stress tensor and a couple stress tensor.

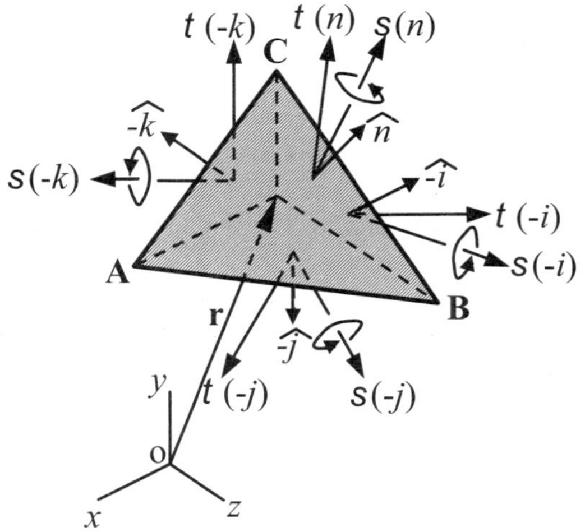


**Figure 2.7.1.** Stresses acting on an infinitesimal tetradedron.

In classical fluid mechanics, all the kinematic parameters are assumed to be determined once the velocity field is specified. Thus, if the velocity field is identically zero, both the linear and angular momentum of all material elements must also be identically zero. However, when a particle has no translation velocity, so that its linear momentum is zero, it may still be spinning about an axis, giving rise to an angular momentum. Thus, angular momentum can still exist even in the absence of linear momentum. In polar mechanics, specifying the velocity field is not sufficient, and additional kinematic measures must be introduced to describe this internal spin.

The most important effect of polar mechanics is the introduction of a size-dependent effect that is not predicted by the classical nonpolar theories. The class of theories in which both the effects of couple stresses and micro-rotation are simultaneously taken into account in a systematic manner are termed micro fluids. Micropolar fluids are the simplest subclass of micro fluids, obtained by restricting the gyration tensor to be skew-symmetric. In micropolar fluids, rigid particles contained in a small volume can rotate about the center of the volume element described by the micro-rotation vector, but neither micro-deformation nor micro-stretch is admissible. This local rotation of the particles is in addition to the usual rigid body motion of the entire volume element. In micropolar fluid theory, the laws of classical continuum mechanics are augmented with additional equations that account for conservation of microinertia moments. Physically, micropolar fluids represent fluids consisting of rigid, randomly oriented particles suspended in a viscous medium, where the deformation of the particles is ignored. The modified momentum, angular momentum and energy equations are:

$$\rho \frac{D\mathbf{U}}{Dt} = \nabla \cdot \tau + \rho \mathbf{f} \tag{2-21}$$

$$\rho I \frac{D\mathbf{\Omega}}{Dt} = \nabla \cdot \sigma + \rho \mathbf{g} + \tau_x \tag{2-22}$$

$$\rho c_p \frac{DT}{Dt} = k\nabla^2 T + \tau : (\nabla \mathbf{U}) + \sigma : (\nabla \mathbf{\Omega}) - \tau_x \cdot \mathbf{\Omega} \tag{2-23}$$

where $\mathbf{\Omega}$ is the micro-rotation vector and $I$ is the associated micro-inertia coefficient; **f** and **g** are the body and couple force vector per unit mass; $\tau$ and $\sigma$ are the stress and couple-stress tensors. $\tau : (\nabla\mathbf{U})$ is the dyadic notation for $\tau_{ji}U_{i,j}$, the scalar product of $\tau$ and $\nabla\mathbf{U}$. If $\sigma$=0 and $\mathbf{g}$=$\mathbf{\Omega}$=0, then the stress tensor $\tau$ reduces to the classical symmetric stress tensor, and the governing equations reduce to the classical model (Lukaszewicz, 1999).

In micropolar fluids, the motion is affected by (i) viscous action, measure by the shear viscosity coefficient $\mu$, (ii) microrotation, measured by the vortex viscosity coefficient $\kappa$, and (iii) the effect of couple stresses, measured by spin-gradient viscosity coefficient $\gamma$. These are material properties that give rise to a length scale $l$:

$$l^2 = \frac{\gamma}{\mu}\left(\frac{\mu + \kappa}{\kappa}\right) \tag{2-24}$$

Although the micropolar fluid theory has been applied to numerous problems, the primary drawback of this anlysis is the unknown viscosity coefficients $\gamma$ and $\kappa$ and, consequently, the polarity length scale $l$. Hence, further work must be done in quantifying the microstructural parameters before the theory van be fully utilized.

### 2.8 Laminar and turbulent flow

The two most important parameters characterizing a fluid, either gas or liquid, are density and viscosity. They are arranged in a non-dimensional form, Reynolds number, defined as:

$$Re = \frac{\rho UL}{\mu} \tag{2-25}$$

which is used to distinguish not only between inviscid and viscous but also between laminar and turbulent flow.

**Table 2.8.1.** Flow classification according to Reynolds number.

| | |
|---|---|
| $Re \to \infty$ | Inviscid Flow |
| $Re > 1$ | Viscous Flow |
| $Re \sim 1$ | Creeping Flow |

For pipe flow (internal flow) with Reynolds number smaller than 2300, the flow is laminar; the flow becomes turbulent if the Reynolds number is larger than 2300. For the boundary layer flow (external flow), the critical Reynolds number is $5 \cdot 10^5$. The boundary layer is laminar if the Reynolds number is smaller than $5 \cdot 10^5$, as the Reynolds number increase transition to turbulence occurs.

# 3 Pressure-Driven Gas Flow in Microchannels

## 3.1 Slip flow in microscale

Considering air flow, the mean free path is about 0.065μm at atmospheric pressure, through a microchannel about 1μm in height; the resulting Knudsen number is on the order of 0.05. Flow with a Knudsen number larger than 0.001 is classified to be in the slip regime. In the slip flow regime, the continuum flow model is still valid for the calculation of the flow properties away from solid boundaries. However, the boundary conditions have to be modified to account for the incomplete interaction between the gas molecules and the solid boundaries

**Velocity slip boundary condition:** The velocity at a certain distance $y=\lambda$ away from the wall, using Taylor expansion, can be written as:

$$u\Big|_{y=\lambda} = u\Big|_{y=0} + \lambda \frac{\partial u}{\partial y}\bigg|_{y=0} + \frac{\lambda^2}{2}\frac{\partial^2 u}{\partial y^2}\bigg|_{y=0} + O(\lambda^3) \tag{3-1}$$

At the wall, $y=0$, the velocity is the average speed of incoming and outgoing molecules such that:

$$u\Big|_{y=0} = u_0 = \frac{1}{2}\{reflected\} + \frac{1}{2}\{incident\} \tag{3-2}$$

For diffuse reflection boundary condition ($\sigma$): $u=0$, while for specular reflection boundary condition ($1-\sigma$): $u=u|_{y=\lambda}$; here, $\sigma=1$ means full accommodation and $\sigma=0$ means zero accommodation. Hence, retaining only the first order term in the Taylor expansion (linear approximation), the velocity at the wall can be written as:

$$u_0 = u\Big|_{y=0} = \frac{1}{2}\left\{\sigma \cdot 0 + (1-\sigma)\left(u_0 + \lambda \frac{\partial u}{\partial y}\bigg|_{y=0}\right)\right\} + \frac{1}{2}\left\{u_0 + \lambda \frac{\partial u}{\partial y}\bigg|_{y=0}\right\} \tag{3-3}$$

Re-arrangement of the terms leads to:

$$u_0 = \frac{1-\sigma}{2}\left( u_0 + \lambda \frac{\partial u}{\partial y}\bigg|_{y=0} \right) + \frac{1}{2}\left( u_0 + \lambda \frac{\partial u}{\partial y}\bigg|_{y=0} \right) \tag{3-4}$$

$$u_0 = \frac{2-\sigma}{2} u_0 + \frac{2-\sigma}{2} \lambda \frac{\partial u}{\partial y}\bigg|_{y=0} \tag{3-5}$$

$$u_0 = \frac{2-\sigma}{\sigma} \lambda \frac{\partial u}{\partial y}\bigg|_{y=0} \tag{3-6}$$

and if $\sigma=1$:

$$u_0 = \lambda \frac{\partial u}{\partial y}\bigg|_{y=0} \tag{3-7}$$

### 3.2 Gas flow in a straight and uniform microchannel

**Theoretical model (simple integral approach):** For a steady 2-D flow with $KnMa \ll 1$,

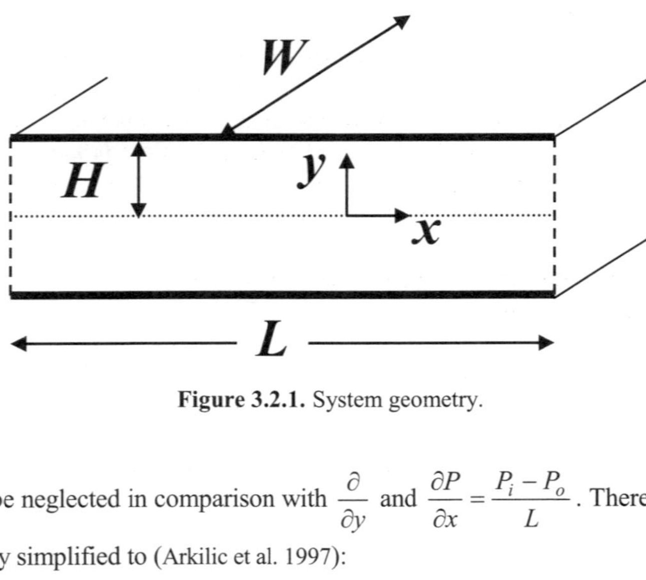


**Figure 3.2.1.** System geometry.

$\frac{\partial}{\partial t}, \frac{\partial}{\partial z}, \frac{\partial}{\partial x}$ can be neglected in comparison with $\frac{\partial}{\partial y}$ and $\frac{\partial P}{\partial x} = \frac{P_i - P_o}{L}$. Therefore the governing equation is greatly simplified to (Arkilic et al. 1997):

$$\frac{\partial^2 u}{\partial y^2} = \frac{1}{\mu}\frac{\partial P}{\partial x} \tag{3-8}$$

Double integration with respect to y-coordinate yields,

$$\frac{\partial u}{\partial y} = \frac{1}{\mu}\frac{\partial P}{\partial x} y + C_1 \tag{3-9}$$

and,

$$u = \frac{1}{2\mu}\frac{\partial P}{\partial x}y^2 + C_1 y + C_2 \tag{3-10}$$

Applying the symmetry boundary condition, $\left.\frac{\partial u}{\partial y}\right|_{y=0} = 0$, leads to $C_1$=0. $C_2$ is found by utilizing the velocity slip boundary condition, $u = -\lambda \left.\frac{\partial u}{\partial y}\right|_{y=\pm H}$, as follows:

$$C_2 = -\frac{H^2}{2\mu}\frac{\partial P}{\partial x}(1 + 4Kn) \tag{3-11}$$

where $Kn = \frac{\lambda}{2H}$. Finally, the velocity profile is given by:

$$u(y) = -\frac{H^2}{2\mu}\frac{\partial P}{\partial x}\left[1 - \left(\frac{y}{H}\right)^2 + 4Kn\right] \tag{3-12}$$

The mass flow rate can be calculated as follows:

$$Q_m = 2\int_0^H \rho(x)u(y)W dy \tag{3-13}$$

$$Q_m = 2\int_0^H \frac{P}{RT}u(y)W dy = -\frac{2WH^3}{3\mu RT}P\frac{\partial P}{\partial x}(P + 6KnP) \tag{3-14}$$

$$Kn(x)P(x) = \frac{\lambda(x)}{2H}P(x) = \frac{\mu}{2H}\left(\frac{\pi RT}{2}\right)^{1/2} = const \tag{3-15}$$

$$Q_m = -\frac{2WH^3}{3\mu RT}\frac{\partial P}{\partial x}\left[P(x) + 6Kn_o P_o\right] \tag{3-16}$$

$$\int_0^x Q_m dx' = -\frac{2WH^3}{3\mu RT}\int_0^x \left[P\frac{\partial P}{\partial x} + 6Kn_o P_o\frac{\partial P}{\partial x}\right]dx' = -\frac{2WH^3}{3\mu RT}\left[\frac{P^2}{2} + 6Kn_o P_o P\right]_0^x \tag{3-17}$$

$$Q_m \cdot x = \frac{WH^3}{3\mu RT}\left[P_i^2 - P^2(x) + 12Kn_o P_o P_i - 12Kn_o P_o P(x)\right] \tag{3-18}$$

$$Q_m = \frac{WH^3 P_o^2}{3\mu RTL}\left[\left(\frac{P_i}{P_o}\right)^2 - 1 + 12Kn_o\left(\frac{P_i}{P_o} - 1\right)\right] \tag{3-19}$$

The mass flow rate at any given point is equal to the outlet mass flow rate, hence:

$$P_i^2 - P^2(x) + 12Kn_oP_oP_i - 12Kn_oP_oP(x) = \frac{x}{L}\left(P_i^2 - P_o^2 + 12Kn_oP_oP_i - 12Kn_oP_o^2\right) \tag{3-20}$$

The streamwise pressure distribution can be obtained by re-arranging Eq. 3-20:

$$\frac{P(x)}{P_o} = -6Kn_o + \left\{\left(6Kn_o + \frac{P_i}{P_o}\right)^2 - \left[\left(\frac{P_i^2}{P_o^2} - 1\right) + 12Kn_o\left(\frac{P_i}{P_o} - 1\right)\right]\frac{x}{L}\right\}^{1/2} \tag{3-21}$$

Using Eq. 3-8 as the governing equation in the simplified model necessarily means that the non-linear convective term is neglected and the velocity profile is presume to be parabolic. A more comprehensive model, in which these restrictions are relaxed, is given by the following continuity and momentum equations:

$$\frac{\partial(\rho u)}{\partial x} = 0 \tag{3-22}$$

$$\rho u\frac{\partial u}{\partial x} = -\frac{\partial P}{\partial x} + \mu\frac{1}{y^m}\frac{\partial}{\partial y}\left(y^m\frac{\partial u}{\partial y}\right) + \mu\left(\frac{4}{3} + \frac{\eta}{\mu}\right)\frac{\partial^2 u}{\partial x^2} \tag{3-23}$$

$$\mu\left(\frac{1}{3} + \frac{\eta}{\mu}\right)\frac{\partial^2 u}{\partial x \partial y} = \frac{\partial P}{\partial y} = 0 \tag{3-24}$$

An analytical solution, using the perturbation expansion method, accounting for both acceleration and non-parabolic velocity profile has been presented by Zohar et al. (2002).

### 3.3 Gas flow in straight but non-uniform microchannels

**Expansion or contraction sections:** Microchannels, with expansion/contraction sections, and pressure microsensors were surface micromachined on silicon substrates. The dimensions of a wide channel are about 40μm×1μm×2000μm, and those of a narrow channel are about 20μm×1μm×2000μm, as shown in Figure 3.3.1 (a) and (b). The narrow and wide channels in each pair are connected via a transition section with an included angle of 5°, 15°, 90° or 180°. 15° and 180° corresponding close-up pictures are shown in Figures 3.3.1 (a) and (b). Pressure sensors are sparsely distributed along the straight sections of the microchannel for reference, and densely around the transition sections. Each sensor is based on polycrystalline silicon piezoresistors arranged in a Wheatstone bridge configuration and attached to a membrane, 100μm×100μm in size. Any pressure difference across the membrane results in its deflection, thus straining the piezoresistors and changing their resistance. The voltage change due to the resistance change is recorded, and the pressure difference can be recovered using calibration curves.

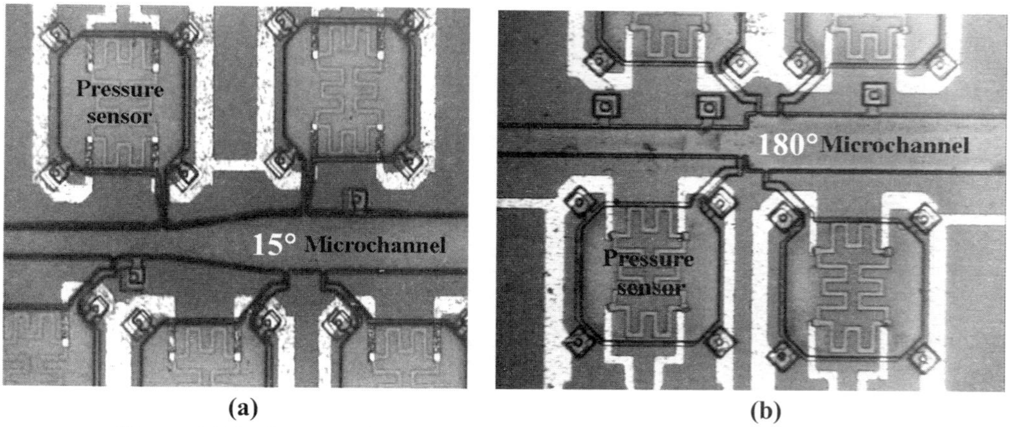


**Figure 3.3.1.** Close-up pictures of transition sections with (a) 15°, and (b) 180°.

**Theoretical mass flow rate under fully attached condition in a series of channels:** Assuming the flow at the inlet and outlet section to be equivalent to a separate channel flow, the inlet and outlet mass flow rate can be written in terms of the inlet/outlet pressure and the pressure at the transition section, $P_{tr}$, as follows:

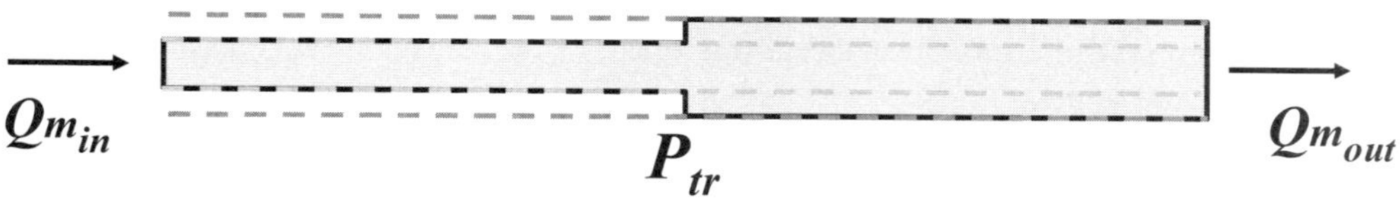


**Figure 3.3.2.** Physical model of theoretical calculation of mass flow rate for a microchannel pair connected in series assuming fully attached flow.

$$Qm_{in} = \frac{h^3 w_{in} P_{tr}^{\ 2}}{24\, RTL\, \mu}\left[\left(\left(\frac{P_{in}}{P_{tr}}\right)^2 - 1\right) + 12\, Kn_{tr}\left(\frac{P_{in}}{P_{tr}} - 1\right)\right] \tag{3-25}$$

and

$$Qm_{out} = \frac{h^3 w_{out} P_{out}^{\ 2}}{24 RTL \mu}\left[\left(\left(\frac{P_{tr}}{P_{out}}\right)^2 - 1\right) + 12 Kn_{out}\left(\frac{P_{tr}}{P_{out}} - 1\right)\right] \tag{3-26}$$

Since the mass flow rate along the entire channel from inlet to outlet is conserved, the inlet and outlet flow rates are equal. Thus, combining Eq'ns 3-25 and 3-26 yields one equation with one unknown, $P_{tr}$. Solving the algebraic equation to obtain $P_{tr}$ allows its substitution into either Eq. 3-25 or Eq. 3-26 to calculate the flow rate through the microchannel pair connected via the transition section.

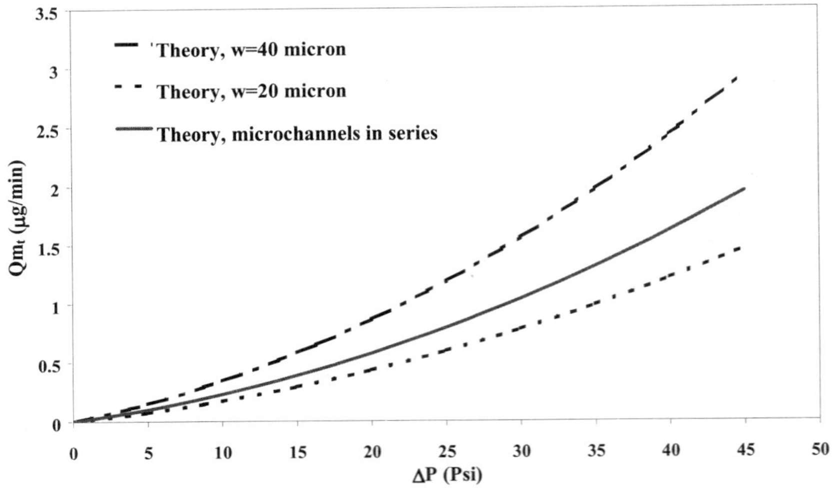


**Figure 3.3.3.** Analytical results of theoretical calculation of mass flow rate for a microchannel pair connected in series assuming fully attached flow.

A comparison between the flow rate through various devices in either contraction or expansion mode is shown in Figures 3.3.4 and 3.3.5, respectively. In each mode, the flow rate decreases monotonically with increasing included angle of the transition section over the entire pressure-drop range tested. Such consistent results indicate that it is more likely that this behavior is not a result of inherent random system errors but rather a true effect of the transition section on the gas flow through the microchannel.

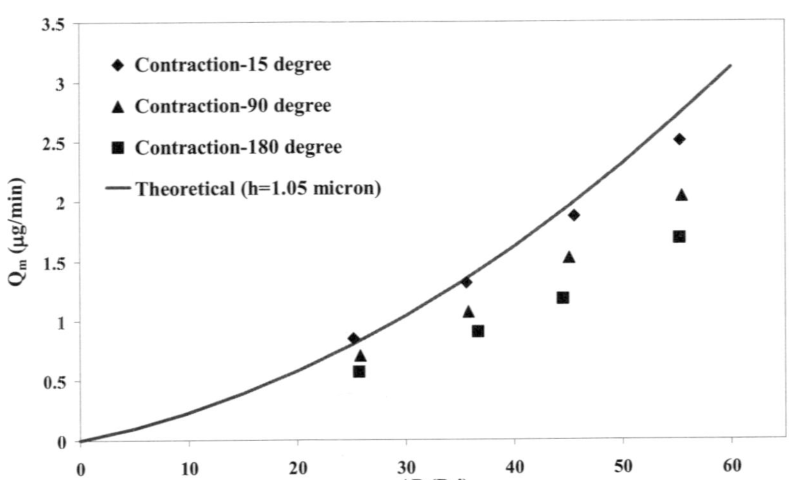


**Figure 3.3.4.** A comparison between the measured mass flow rate for the three included angles in contraction.

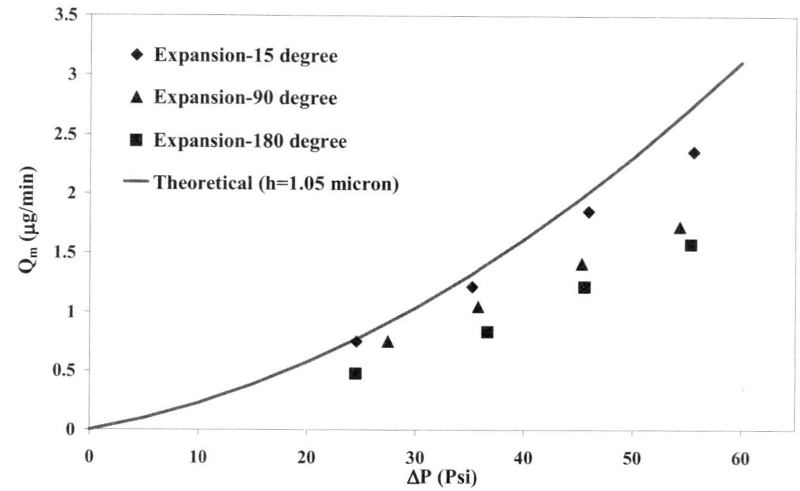


**Figure 3.3.5.** A comparison between the measured mass flow rate for the four included angles in expansion mode.

**Measure streamwise pressure distributions:** Pressure measurements were collected for all devices under three flow-rate levels for either the contraction or the expansion mode. The results for the contraction and expansion mode are summarized respectively in Figures 3.3.6 and 3.3.7 for the 90°-included angle device, of which all the sensors functioned properly. As expected the pressure gradient along the narrow section, whether it is next to the inlet or the outlet, is higher than the gradient along the wide section due to the higher frictional loss.

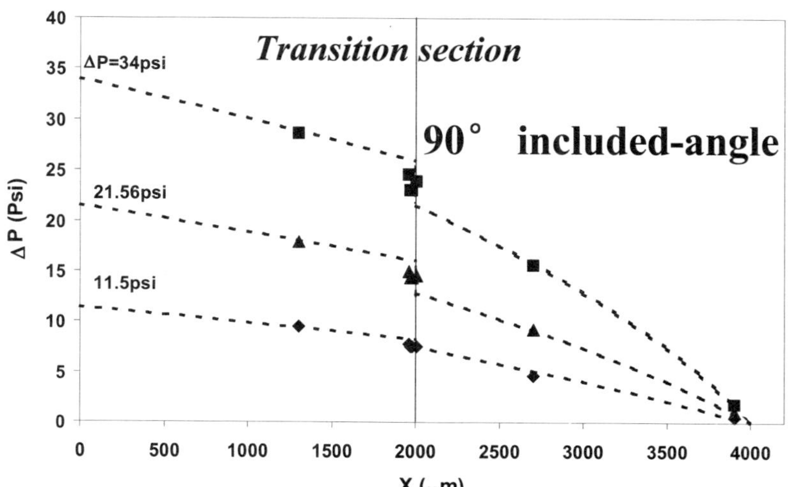


**Figure 3.3.6.** Streamwise pressure distributions along the 90°-included angle device in the contraction.

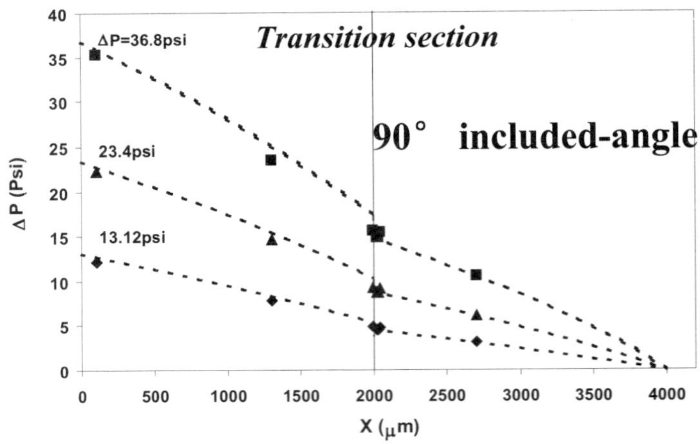


**Figure 3.3.7.** Streamwise pressure distributions along the 90°-included angle device in the expansion.

**Constrictions:** Several microdevices were surface micromachined on silicon wafers. A constriction was created at the center of each microchannel, 40μm×1μm×4000μm in dimensions. Pressure sensors are distributed sparsely along the straight sections of the microchannel for reference and densely around the constriction element. The sensors are based on piezoresistive sensing elements, arranged in a Wheatstone bridge configuration and attached to a membrane, 100μm×100μm in size. The constriction was realized by two symmetric obstructions, each 10μm wide with varying length. A few devices were fabricated with the channel width $d$=10, 20, 28 and 34μm at the constriction. Close-up pictures of the constrictions with the smallest and largest gaps tested in this work are shown in Figure 3.3.8 (a) and (b), respectively.

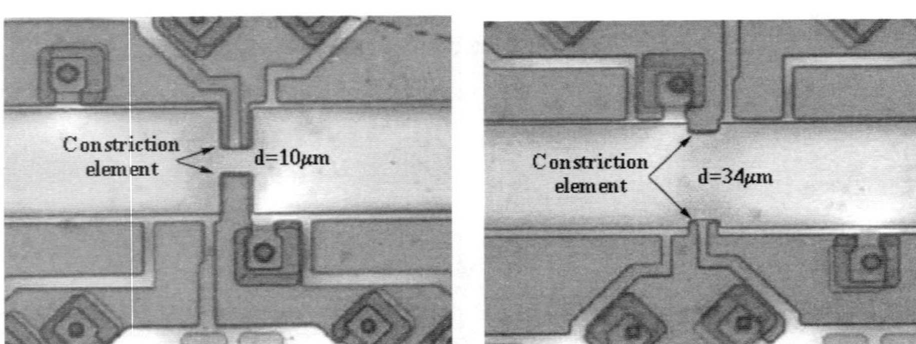


(a) (b)

**Figure 3.3.8.** Close-up pictures of the constriction elements in two of the tested devices with constriction-gap width of (a) $d$=10μm, and (b) $d$=34μm.

**Constriction-gap width effect:** Nitrogen flow rate, $Q_m$, was measured for all devices as a function of the pressure drop, $\Delta P$, and the results are summarized in Figure 3.3.9. The solid curve is based on theoretical calculations as equation 3-19 for a straight microchannel, 40μm×0.95μm×4000μm in dimensions. Due to the presence of the obstructions, the measured mass flow rate in all devices with constrictions is smaller than the calculated rate for the straight channel in the range of Δ *P* tested. The flow rate in a straight, uniform channel is inversely proportional to the channel length. The maximum additional length in the channel with the narrowest constriction is about 60μm, for *d*=10μm, while the channel length is around 4000μm. Hence, the mass flow rate should decrease by no more than 1.5% of the rate through a straight channel. However, even for the largest constriction gap, *d*=34μm, the measured values are about 15% smaller than the uniform channel flow rate. The measured flow rate for the smallest constriction gap, *d*=10μm, is significantly lower, about 55% of the uniform channel rate.

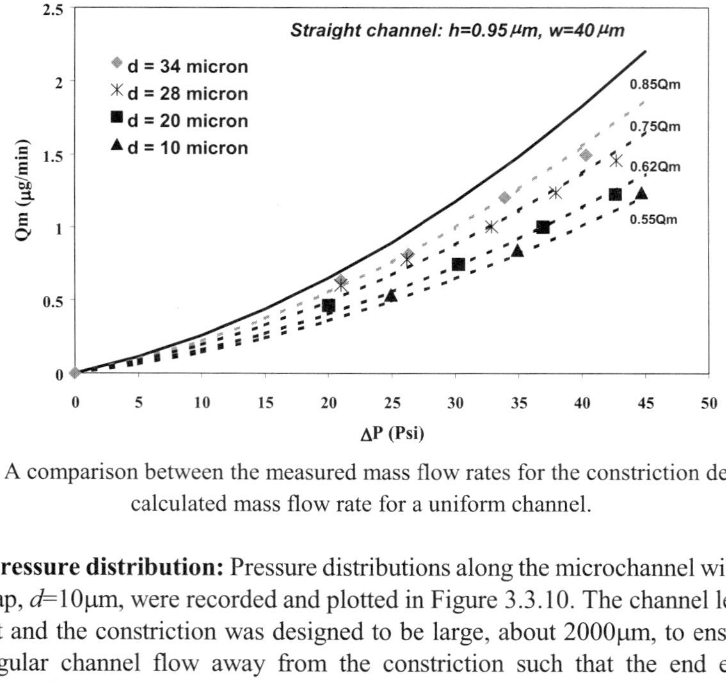


**Figure 3.3.9.** A comparison between the measured mass flow rates for the constriction devices and the calculated mass flow rate for a uniform channel.

**Streamwise pressure distribution:** Pressure distributions along the microchannel with the smallest constriction gap, *d*=10μm, were recorded and plotted in Figure 3.3.10. The channel length between the inlet/outlet and the constriction was designed to be large, about 2000μm, to ensure the development of regular channel flow away from the constriction such that the end effects can be neglected. Extrapolating the measured pressure distribution towards the constriction element, the outlet pressure of the upstream section, $Pu_o$, and the inlet pressure of the downstream section, $Pd_i$, can be estimated. The pressure distribution is then calculated for each section, based on the same analytical model used for deriving Eq. 3-21, as follows:

$$\frac{P(x)}{Pu_o} = -6Kn_{uo} + \left\{\left(6Kn_{uo} + \frac{P_i}{Pu_o}\right)^2 - \left[\left(\frac{P_i^2}{Pu_o^2} - 1\right) + 12Kn_{uo}\left(\frac{P_i}{Pu_o} - 1\right)\right]\left(\frac{x_i}{L/2}\right)\right\}^{1/2} \quad (3\text{-}27)$$

$$\frac{P(x)}{P_o} = -6Kn_o + \left\{\left(6Kn_o + \frac{Pd_i}{P_o}\right)^2 - \left[\left(\frac{Pd_i^2}{P_o^2} - 1\right) + 12Kn_o\left(\frac{Pd_i}{P_o} - 1\right)\right]\left(\frac{x_o}{L/2}\right)\right\}^{1/2} \quad (3\text{-}28)$$

where $x_i$ is the streamwise distance from the channel inlet, $x_o$ is the streamwise distance from the constriction element at the channel center, and $Kn_{uo}=P_oKn_o/P_{uo}$ is the Knudsen number at the upstream-section outlet. The calculated nonlinear pressure distributions outside the constriction region agree reasonably well with the measured values as demonstrated in Figure 3.3.10. Small deviations, especially for the high-pressure drop case, could be the result of the bend effect. This suggests that the flow development in the straight sections between the inlet/outlet and the constriction element is similar to that in a straight microchannel.

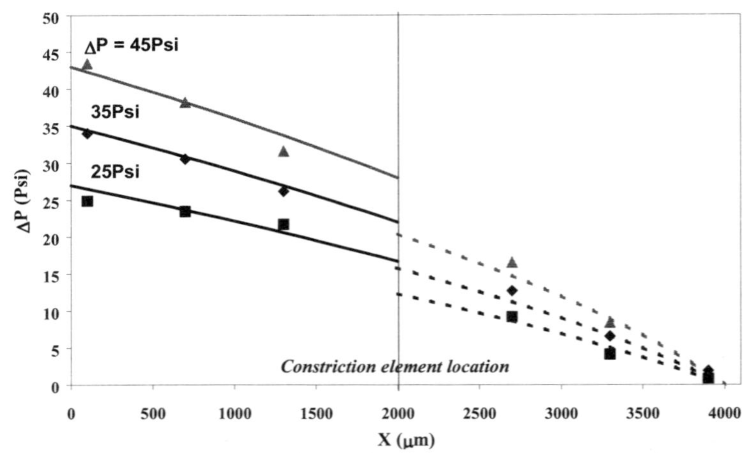


**Figure 3.3.10.** Streamwise pressure distributions along the 10μm-constriction microchannel.

**Pressure loss:** The pressure loss due to the transition section, estimated from the measurements presented in Figure 3.3.10, is plotted in Figure 3.3.11. Based on the estimated static pressure either upstream or downstream of the constriction element, the flow-rate calculations (based on equation 3-19) can be corrected by using either one of the following relationships:

$$Q_m = \frac{H^3DPu_o^2}{24RT(L/2)\mu}\left[\left(\frac{P_i}{Pu_o}\right)^2 - 1 + 12Kn_{uo}\left(\frac{P_i}{Pu_o} - 1\right)\right] \tag{3-29}$$

$$Q_m = \frac{H^3DP_o^2}{24RT(L/2)\mu}\left[\left(\frac{Pd_i}{P_o}\right)^2 - 1 + 12Kn_o\left(\frac{Pd_i}{P_o} - 1\right)\right] \tag{3-30}$$

The corrected mass flow rates for all devices are compared in Figure 3.3.12 with the measured values. The theoretical predictions are within 5% of the experimental data. This reasonable agreement confirms that the additional loss cannot be attributed to additional friction loss due to the larger surface area of the constriction device.

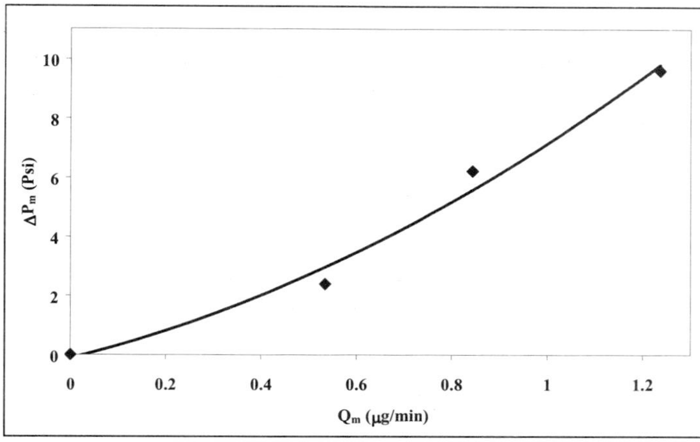


**Figure 3.3.11.** Pressure loss plotted again with mass flow rate ($d$=10μm, $\theta$= 90°).

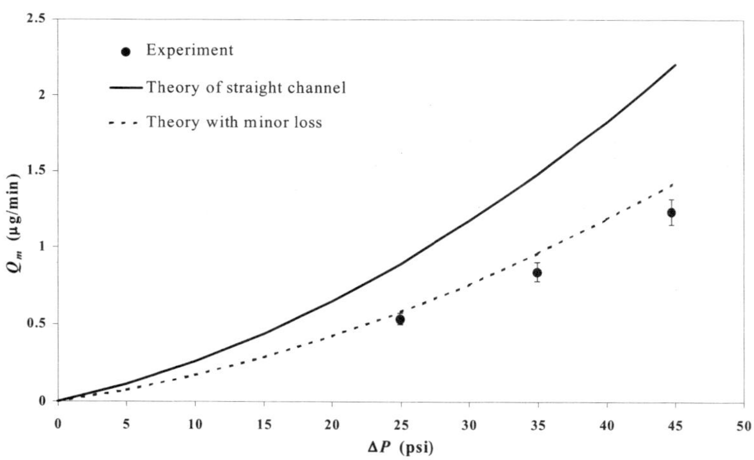


**Figure 3.3.12.** A comparison between the measured and corrected calculations of mass flow rate accounting for the pressure loss due to the 10μm-gap constriction.

**Cavities:** A variety of microchannels with cavities have been designed to study numerous effects. An example picture of a fabricated device is shown in Figure 3.3.13 and the cavity layout is shown in Figure 3.3.14. In one set of microchannels, 4000×40×1.5μm$^3$ in dimensions, a single cavity was designed to examine the effect of its particular configuration on the flow in the microchannel and in the micro cavity. The single cavity, always at the microchannel center, varied between $L_c$= 40 to 160μm in length and between $W_c$= 40 to 800μm in width. Another set of microchannels, 3825×30 ×1.5μm$^3$ in dimensions, was designed to study the number of cavities effect on the flow field. The cavity configuration was kept constant in this set, $L_c$=$W_c$=15μm, while the number of cavities increased from 1 to a maximum of 127 as shown in Figure 3.3.15; with the cavities placed symmetrically with respect to the channel center at equal intervals. Both sets included integrated pressure sensors designed in a Wheat-Stone bridge configuration of 4 piezoresistive elements. The third set of microchannels, $400\times100\times20\mu m^3$ in dimensions, was designed to study the flow pattern with the cavities. A pair of square cavities, varying between $20\times20$ to $70\times70\mu m^2$ in area, was placed at each microchannel center. In all three device sets, the height of the cavities has been designed to be equal to the height of the microchannel.

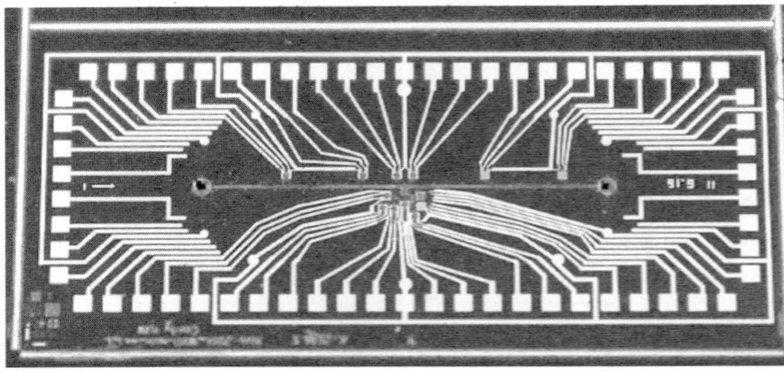

**Figure 3.3.13.** A photograph of a fabricated microdevice.

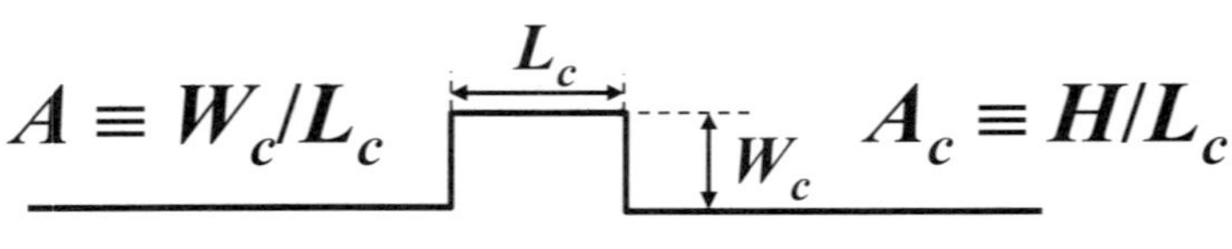


**Figure 3.3.14.** A schematic illustration of a microchannel with a single cavity.

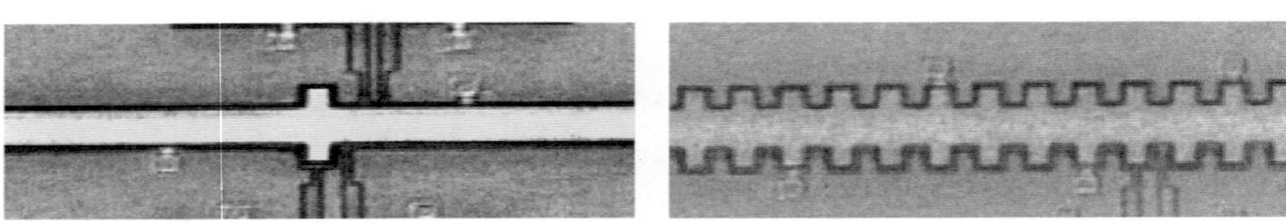

(a) (b)

**Figure 3.3.15.** Pictures of fabricated microchannels with (a) a single cavity, and (b) multi-cavity pairs (127) for gas flows.

**Theoretical analysis (Hele-Shaw flow):** The argon mean free path at atmospheric pressure is 0.104μm; hence, the outlet Knudsen number for a 1.6μm-high microchannel is about $Kn_o$=0.065. Hence, the flow is considered to be in the slip regime; the flow interior is still governed by the continuum equations with the velocity-slip boundary condition. This is the mathematical model used in the numerical computations of gas flows, while no-slip boundary condition is used in liquid flow, utilizing a commercially available finite-volume code CFD-ACE(U) developed by CFD Research Corp. The code verification was done for the case of a uniform microchannel, i.e. no cavities, for which a simple analytical solution has been derived.

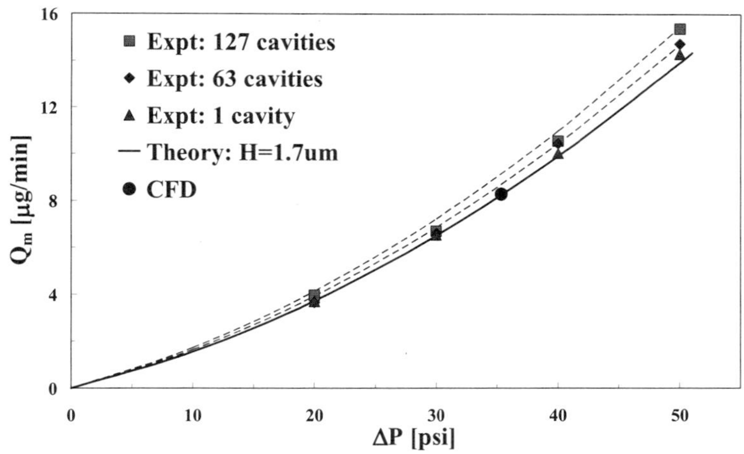


**Figure 3.3.16.** Comparison between experimental results, theory model and CFD result.

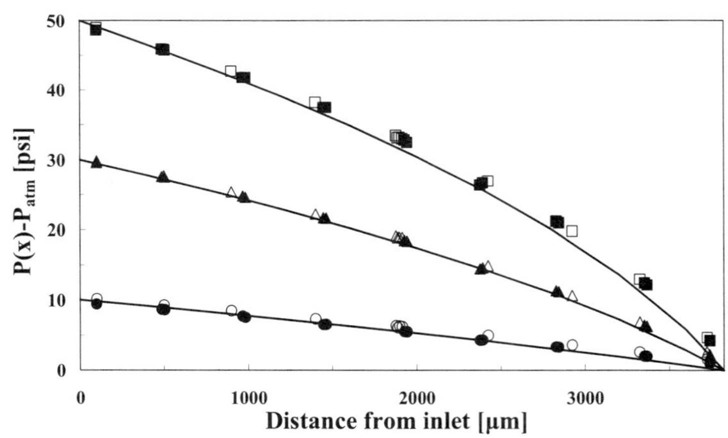


**Figure 3.3.17.** Measured streamwise pressure distributions for various total pressure drops in microchannels with multi-cavity pairs, compared with theoretically calculated distributions in a uniform microchannel. Hollow symbol means single cavity and solid symbol means multi-cavity pairs.

Assuming 2-D, fully-developed flow in a microchannel, the mass flow rate can be estimated using Eq. 3-19. The equation accounts for compressible and slip flow effects but not for non-parabolic velocity and acceleration effects. A comparison between the measured and calculated mass flow rate is summarized in Figure 3.3.16. Pressure losses due to the channel inlet and outlet have also been neglected in the analysis since the channel length is much larger than its height, $L/H>103$. The corresponding streamwise pressure distributions given by Eq. 3-21 are compared with experimental measurements in Figure 3.3.17. The mass flow rate increase almost linearly with the number of cavities as shown in Figure 3.3.18.

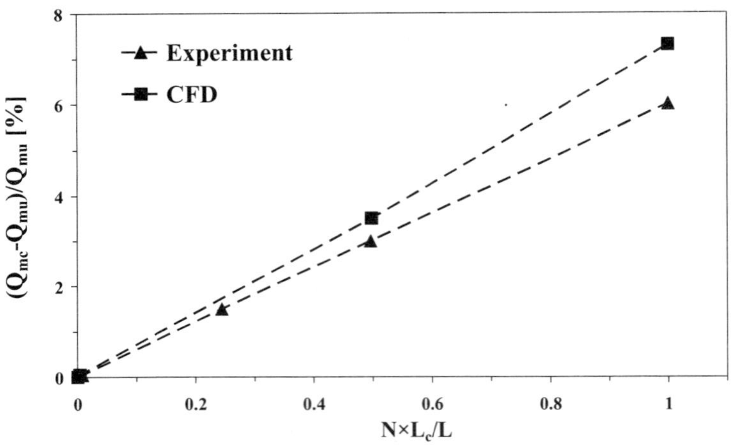


**Figure 3.3.18.** The effect of the number of cavity pairs along a microchannel on the measured and computed mass flow rate.

While the flow in the microchannel is fully-attached, the flow in the cavities can be either separated or attached. It is similar to flow between two parallel plates separated by a narrow gap known as Hele-Shaw flow characterized by two conditions (Riegels, 1938). The first condition is negligible inertia force compared to the friction force, $\rho u(\partial u/\partial x) << \mu(\partial^2 u/\partial y^2)$, leading to a small reduced Reynolds number $Re^*$:

$$Re^* = \frac{\rho U_a L_c}{\mu}\left(\frac{H}{L_c}\right)^2 << 1 \tag{3-34}$$

where $U_a$ is a velocity scale for the cavity flow. This is a well-known parameter often called reduced Reynolds number. The second condition is negligible planar viscous terms compared to the cross-stream term, $(\partial^2 u/\partial x^2) << (\partial^2 u/\partial y^2)$, resulting in a small cavity number, $A_c$:

$$Ac = \left(\frac{H}{L_c}\right)^2 << 1 \tag{3-35}$$

It is important to emphasize that though this ratio appears in the condition for the reduced Reynolds number, it is an independent condition strictly geometric. Although this control parameter has been noted in the literature, its effect has never been explicitly discussed. The cavity flow is driven by the flow in the main microchannel. Therefore, the velocity scale is taken as the average velocity in the microchannel given by: $U_a = Q_m/\rho A$, where $A$ is the channel cross-sectional area. Since $Q_m$ and $A$ are readily available not only in the simulations but also in the experiments, both parameters $Re^*$ and $A_c$ are known for every tested configuration.

The analysis for Hele-Shaw flow revealed two control parameters: $Re^*$ and $A_c$, defined in equations 3-34 and 3-35, respectively. Numerical computations of the cavity flow field resulted in attached or separated flow in the cavity depending on the values of these parameters. In order to support the simulation results, flow visualization experiments have been conducted in which the value of the control parameters varied systematically. The planar velocity vector plot at the cavity mid-height, $y_c = 0$, is shown in Figures 3.3.19(a) to (c) for $Re^* = 3$, 11 and 100, respectively, having the same $A_c$=0.57. Evidently, Figures 3.3.19(a), the flow is fully attached for the smaller reduced Reynolds number, $Re^* = 3$. Increasing the reduced Reynolds number to $Re^* = 11$, Figures 3.3.19(b), results in a separated flow inside the cavity although the height is very small, H~10μm. To underline the independent effect of the cavity number, similar simulations has been repeated for different values of $A_c$ with constant $Re^*$, and the results are respectively shown in Figure 3.3.20(a) to (c). Indeed, the flow is fully attached for the lower cavity number, $A_c = 0.25$, Figure 3.3.20(a). However, even for the same very low reduced Reynolds number, $Re^*$=0.15, flow separation in the cavity occurs if the ratio is increased to $A_c$=0.71 as shown in Figures 3.3.20(b).

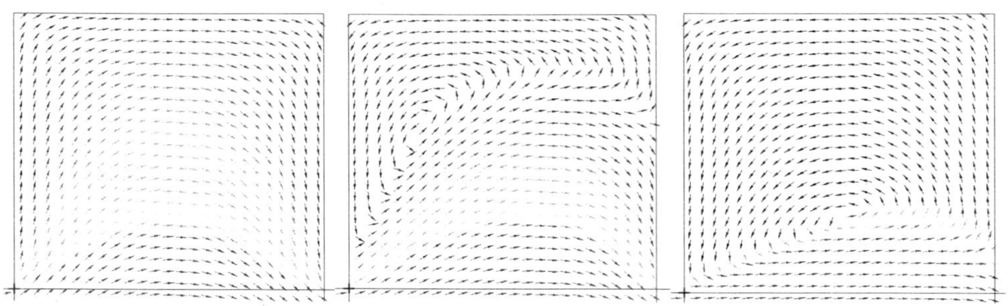

(a) (b) (c)

**Figure 3.3.19.** Numerical computations; gas flow pattern dependence on the reduced Reynolds number for the cavity number with $A_c = 0.57$, (a) $Re^* = 3$, (b) $Re^* = 11$ and (c) $Re^* = 100$.

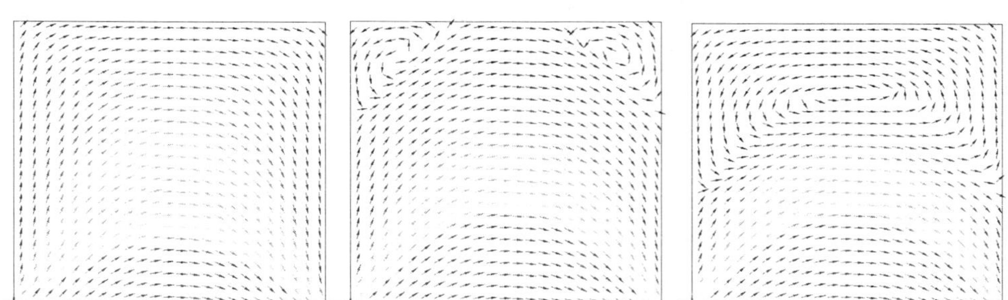

(a) (b) (c)

**Figure 3.3.20.** Numerical computations; gas flow pattern dependence on the cavity number for the reduced Reynolds number with $Re^* = 0.15$, (a) $A_c = 0.57$, (b) $A_c = 0.71$ and (c) $A_c = 0.86$.

The cavity flow pattern is found to undergo transition from attached to separated with increasing either $Re^*$ or $A_c$, and it is interesting to find the critical values for this transition. In the absence of an adequate analytical model, flow patterns for various combinations of $Re^*$ or $A_c$ need to be examined. All the combinations of $Re^*$ and $A_c$ used in this investigation are marked in Figure 3.3.21. Based on the computed velocity vector plot, as in Figure 3.3.19 and Figure 3.3.20, each cavity flow pattern is determined to be either attached or separated and, correspondingly, denoted by either hollow or full symbols. This phase-plot diagram can now be considered as a flow regime map with clear boundaries between different flow patterns. The flow in the cavity is fully attached only if two criteria are simultaneously met: $Re^*<4$ and $A_c<0.327$. Flow separation occurs in the cavity if either $Re^*>10$ or $A_c>0.327$. Whereas the simple analysis required that both $Re^*$ and $A_c$ be very small, the computations provide sharp critical values.

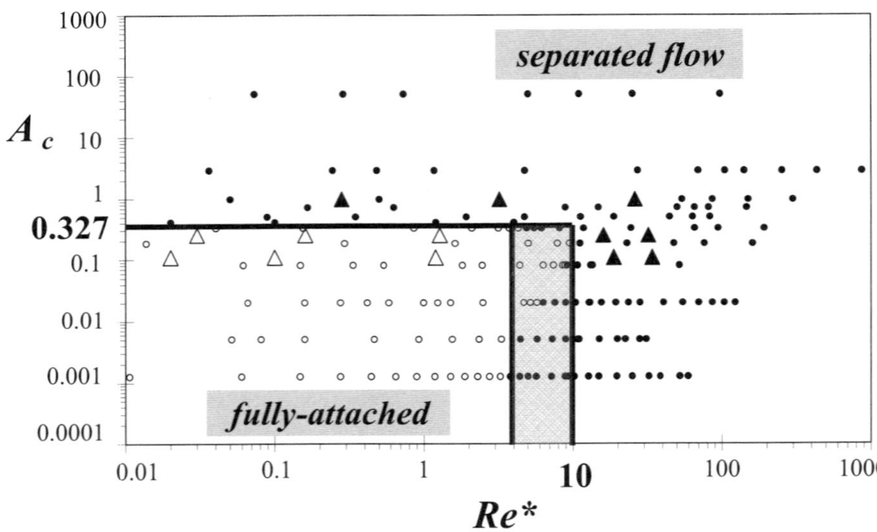


**Figure 3.3.21.** Flow pattern diagram.

### 3.4 Gas flow in non-straight but uniform microchannels

**Bends:** Microdevices with bended microchannels were surface micromachined on silicon substrates. Each device included a microchannel, 20μm×1μm×5810μm in dimensions. All the microchannels were designed with a 90°-turn mid-way between the inlet and outlet to eliminate any possible interaction between end effects and bend effects. Microchannels with a variety of bend configurations were fabricated, and the results pertaining to three types are discussed in this report. Close-up pictures of these bends: miter, curved and double-turn are shown in Figures 3.4.1(a)-(c), respectively. In the miter bend the channel turns sharply at a right angle, while in the curved-bend the channel turns smoothly with a 60μm radius-of-curvature and in the double-turn bend the channel turns sharply twice at 45° with a separation of 50μm between the two bends.

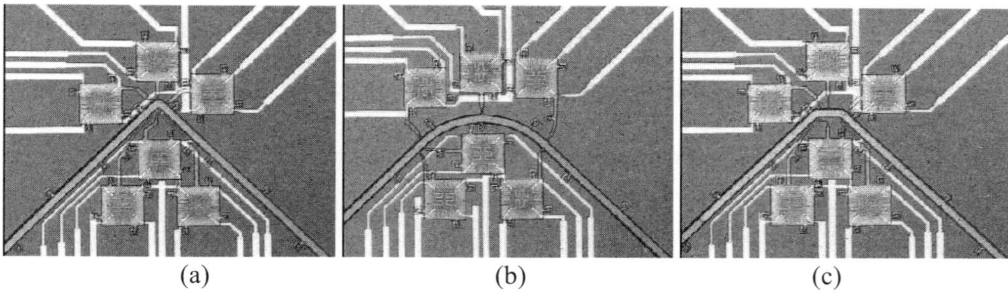


**Figure 3.4.1.** Close-up pictures of 3 different bends located at the center of the 20μm-wide microchannels: (a) miter, (b) curved, and (c) double turn

Mass flow rate of argon, $Q_m$, was measured as a function of the total pressure drop, $\Delta P$, for the channels with the three different bends. The results are compared in Figure 3.4.2 with the calculated flow rate for the straight microchannel. The measured mass flow rates in all the microchannels with bends are smaller than the analytical values. The flow rate through the miter-bend channel is the lowest, about 70%, while the flow rate through either the double turn or the curved-bend channel is about 90% of the calculated flow rate for a straight channel. Since the cross-section is constant, the flow rate through all the devices should be equal to a straight channel flow rate if the flow is fully attached with no separation, such that the only resistance to the flow is due to friction. The lower measured values indicate that the presence of the bends results in additional pressure loss and, subsequently, in reduction of the flow rate. Furthermore, the loss due to one sharp turn of 90°, miter, is larger than the loss due to two sharp turns of 45°. This clearly demonstrates that the overall bend angle, 90° in all the devices, is less important than the detailed geometry of the bend.

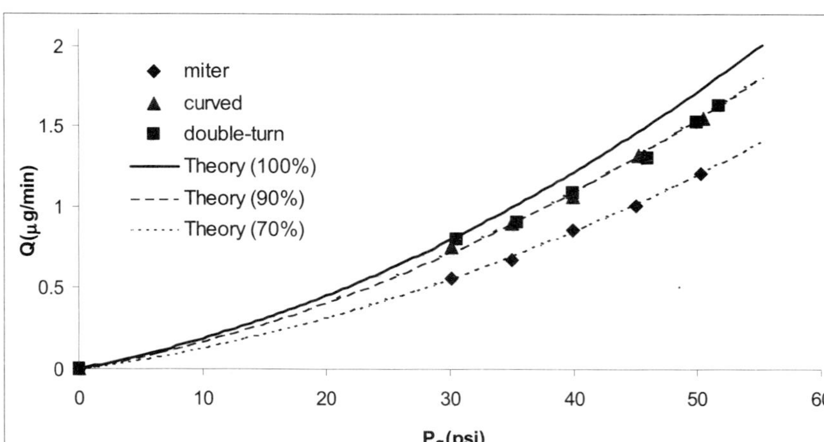


**Figure 3.4.2.** A comparison between the measured mass flow rate for the three microchannels with bends and calculated mass flow rate for a straight microchannel.

In order to investigate the cause of the additional loss, pressure distributions along the three microchannels were recorded. The results for the miter bend are summarized in Figures 3.4.3.

Extrapolating the measured pressure distribution towards the bend, the outlet pressure of the up-stream straight section, $Pu_o$, and the inlet pressure of the downstream straight section, $Pd_i$, can be estimated. The pressure distribution is then calculated as following:

$$\frac{P(x)}{Pu_o} = -6Kn_{uo} + \left\{ \left( 6Kn_{uo} + \frac{P_i}{Pu_o} \right)^2 - \left[ \left( \frac{P_i^2}{Pu_o^2} - 1 \right) + 12Kn_{uo} \left( \frac{P_i}{Pu_o} - 1 \right) \right] \left( \frac{x_i}{L/2} \right) \right\}^{1/2} \quad (3\text{-}36)$$

$$\frac{P(x)}{P_o} = -6Kn_o + \left\{ \left( 6Kn_o + \frac{Pd_i}{P_o} \right)^2 - \left[ \left( \frac{Pd_i^2}{P_o^2} - 1 \right) + 12Kn_o \left( \frac{Pd_i}{P_o} - 1 \right) \right] \left( \frac{x_o}{L/2} \right) \right\}^{1/2} \quad (3\text{-}37)$$

where $x_i$ is the streamwise distance from the channel inlet, $x_o$ is the streamwise distance from the bend at the channel center, and $Kn_{uo}=P_oKn_o/Pu_o$ is the Knudsen number at the upstream-section outlet. The channel length between the inlet/outlet and the bend was designed to be large, about 2900μm, to ensure the development of regular channel flow away from the bend such that the end effects can be neglected. Indeed, the calculated nonlinear pressure distributions outside the bend region, shown in Figure 3.4.3, agree reasonably well with the measured values. This suggests that the flow development in the straight sections between the inlet/outlet and the bend is similar to that in a straight microchannel.

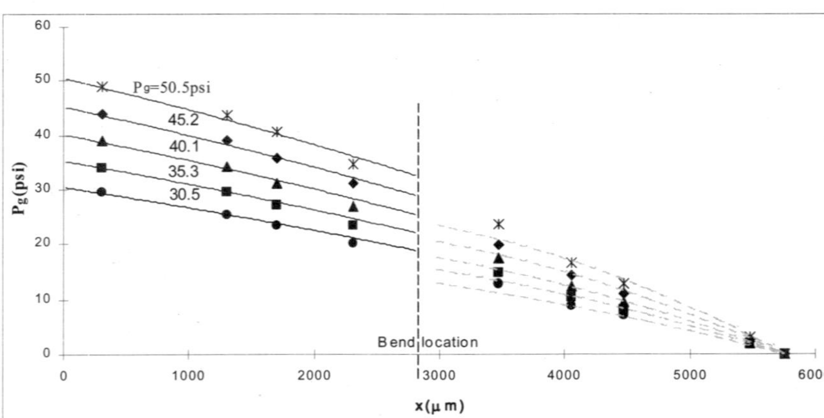


**Figure 3.4.3.** Streamwise pressure distributions along the microchannels with the miter.

A distinct drop in the static pressure across the bends is evident in the streamwise pressure distributions plotted in Figure 3.4.3. This pressure drop can be estimated, $\Delta P = Pu_o - Pd_i$, and is plotted in Figure 3.4.4 as a function of the mass flow rate. The pressure drop increases monotonically with increasing mass flow rate (increasing total pressure drop). Accounting for this additional pressure loss due to the bend, often labeled as a minor loss, the calculated flow rates can be corrected.

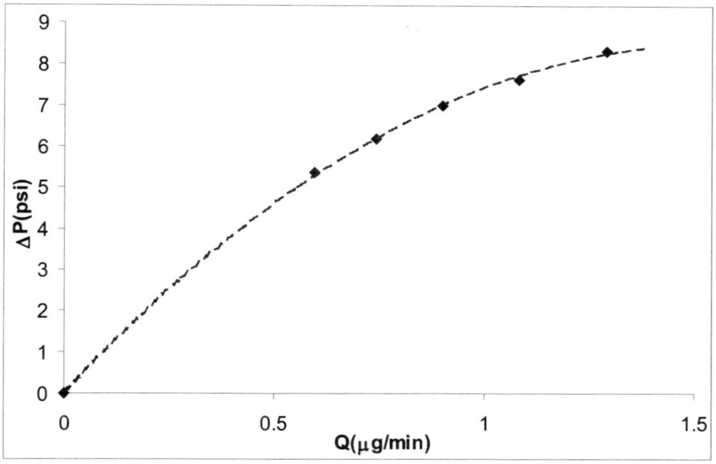


**Figure 3.4.4.** Estimated pressure loss due to the bends as a function of the mass flow rate.

Based on the estimated static pressure either upstream or downstream of the bend, the flow-rate calculations can be corrected by using either one of the following equations:

$$Q_m = \frac{H^3 W P u_o^2}{24RT(L/2)\mu}\left[\left(\frac{P_i}{Pu_o}\right)^2 - 1 + 12Kn_{uo}\left(\frac{P_i}{Pu_o} - 1\right)\right] \tag{3-38}$$

$$Q_m = \frac{H^3 W P_o^2}{24RT(L/2)\mu}\left[\left(\frac{Pd_i}{P_o}\right)^2 - 1 + 12Kn_o\left(\frac{Pd_i}{P_o} - 1\right)\right] \tag{3-39}$$

The corrected mass flow rates for the microchannels with the miter bend is compared in Figure 3.4.5 with the measured values. The theoretical predictions are within 5% of the experimental measurements. This agreement confirms that the additional loss is a direct result of the bend.

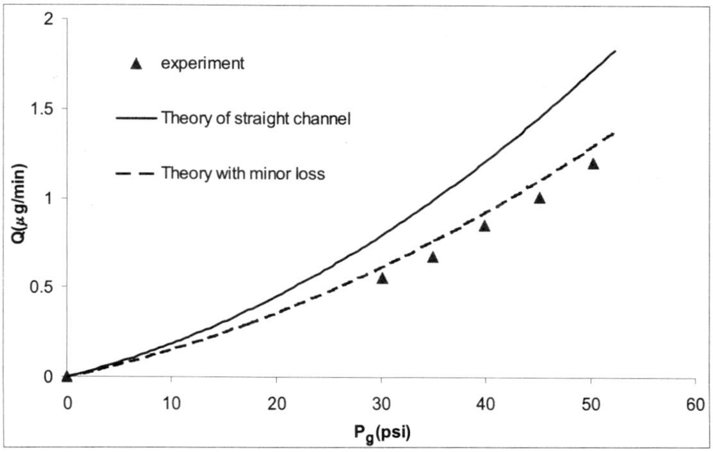


**Figure 3.4.5.** A comparison between the measured and corrected calculations of mass flow rate accounting for pressure loss due to the bends.

# 4 Liquid Flow in MicroSystems

## 4.1 Electric Double Layer (EDL)

Most solid surfaces acquire a surface electric charge when brought into contact with an electrolyte (liquid). Mechanisms for the spontaneous charging of surface layers include:

- differential adsorption of ions from an electrolyte onto solid surfaces (e.g., ionic surfactants);
- differential solution of ions from the surface to the electrolyte;
- deprotonation/ionization of surface groups.

In microfluidic electrokinetic systems, the most common of the above mechanisms is the deprotonation of surface groups on the surface of materials like silica, glass, acrylic and polyester.

**Charge separation at interfaces.** Two phases in contact will spontaneously develop a separation of charge as shown in Figure 4.1.1.

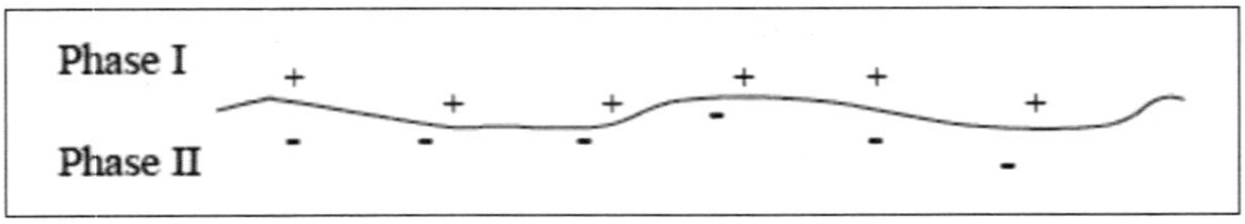


**Figure 4.1.1.** Charge separation of interfaces.

The most important mechanism for charge separation in micro fluidic systems is the ionization of surface groups. For example:

- Carboxyl groups: $COOH \Leftrightarrow COO^- + H^+$;
- Amine groups: $NH_2 + H^+ \Leftrightarrow NH_3^+$.

One of the most applicable reactions to electrokinetics systems etched in glass substrates (or fused silica substrates) is the deprotonation of silanol groups, in which the generated surface charge density is determined by the number of deprotonation of surface silanol groups (SiOH). As a result, the magnitude of the net surface charge density at the liquid/solid interface is a strong function of the local pH value. The equilibrium reaction associated with this deprotonation process can be represented as:

- $SiOH \Leftrightarrow SiO^- + H^+$.

Several characteristics of glass or silicon dioxide film are listed as the following:

- an amorphous solid;
- has the structure of a liquid;
- has no distinct melting or freezing temperature, but only glass transition temperature;
- contains at least 50% silica (a 'polymorphic' material).

The wet and dry states of a properly treated glass/oxide surface can be described as:

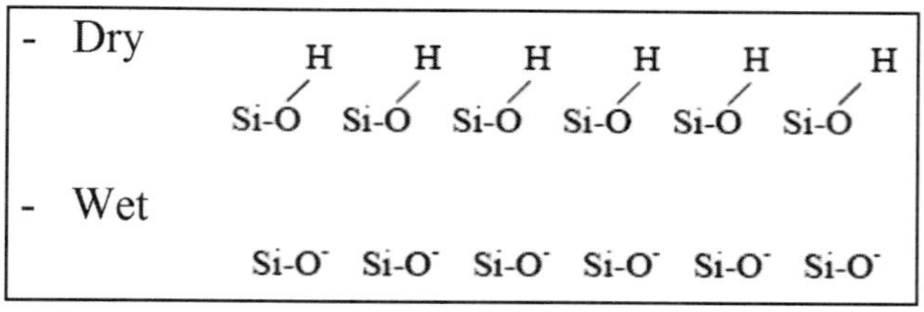


**Figure 4.1.2.** Wet and dry states of a treated glass/silicon dioxide surface.

**Gouy-Chapman model.** According to the charge distribution in Gouy-Chapman theory (Bard & Faulkner, 1979), a balance of force in the ions developed between electromigration, of which moves ions to the wall and the diffusion, which drives ions away from the wall. All the ions in these laminae are assumed to be in thermal equilibrium as stated in the 1-D equation:

$$-\frac{z_i e}{kT}\frac{d\varphi}{dy} = \frac{1}{n_i}\frac{dn_i}{dy}, \tag{4-1}$$

where $z_i$ is the valence number, $e$ the elementary charge, $T$ the temperature, $n_i$ the ion concentration, $\varphi$ the electric potential measured respect to the bulk solution and $k$ the Boltzmann constant. Under the boundary conditions:

$$\varphi \rightarrow 0 \quad \text{at} \quad y \rightarrow \infty, \tag{4-2a}$$

where the bulk is assumed to be neutral charged, and,

$$n^i \rightarrow n^i_0 \quad \text{at} \quad y \rightarrow \infty, \tag{4-2b}$$

$n^i_0$ denotes concentration far from wall. In this case, as shown in Figure 4.1.3, the ion distribution satisfies Boltzmann statistics:

$$n_i(y) = n_i^0 \exp\left(-\frac{z_i e\varphi(y)}{kT}\right). \tag{4-3}$$

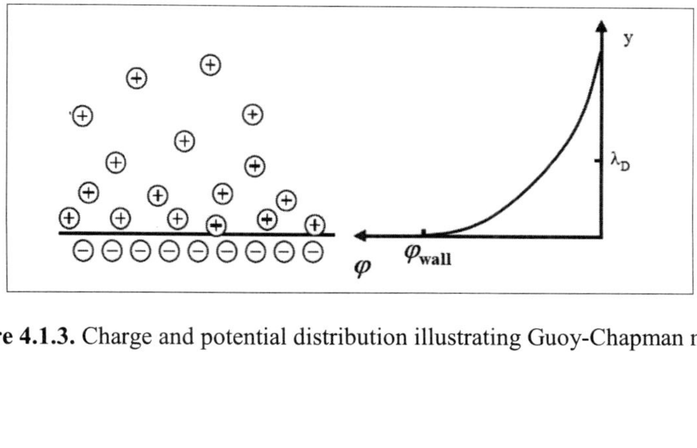


**Figure 4.1.3.** Charge and potential distribution illustrating Guoy-Chapman model.

For the case of general electrolyte, the net charge density ($\rho_e$) in the EDL is distributed as:

$$\rho_e = \sum_{i=1}^{N} z_i e n_i \,. \tag{4-4}$$

The potential due to the charge near the wall obeys Poisson's equation:

$$\frac{d^2\varphi(y)}{dy^2} = -\frac{\rho_e(y)}{\varepsilon\varepsilon_o}, \tag{4-5}$$

where $\varepsilon$ represents the relative permittivity and $\varepsilon_o$ is the absolute permittivity of air. Finally, the charge density can be expressed in terms of ion concentration yielding the Poisson-Boltzmann equation:

$$\frac{d^2\varphi}{dy^2} = \frac{-e}{\varepsilon\varepsilon_0}\sum_{i=1}^{N} z_i n_i^0 \exp\left(-\frac{z_i e\varphi}{kT}\right) \tag{4-6}$$

**Debye-Huckel approximation.** The Poisson-Boltzmann equation is intrinsically non-linear. Debye-Huckel theory allows an important simplification. Considering the limit:

$$z_i e\varphi(y)/kT << 1, \tag{4-7}$$

the Poisson-Boltzmann equation can be simplified as Debye-Huckel equation:

$$\frac{d^2\varphi}{dy^2} = \frac{\varphi}{\lambda_D^2}, \tag{4-8}$$

where the Debye length, $\lambda_D$, is defined as:

$$\lambda_D \equiv \left(\varepsilon kT / e^2 \sum_{i=1}^{N} z_i^2 n_i^0\right)^{1/2}. \tag{4-9}$$

The Debye-Huckel equation is linear in nature, with the boundary conditions:

$$\varphi \to \varphi_w \quad \text{at} \quad y \to \infty, \tag{4-10a}$$

assuming finite electrical potential at the wall ( $\varphi_w$ )

$$\frac{d\varphi}{dy} \to 0 \quad \text{at} \quad y \to \infty, \tag{4-10b}$$

as the ion concentration becomes uniform far from the charged walls. This results the solution:

$$\varphi = \varphi_w \exp\left(-\frac{y}{\lambda_D}\right). \tag{4-11}$$

The electric potential decays exponentially away from the wall. The Debye length ($\lambda_D$) describes the location of the 1/e decay.

**Stern layer.** The Gouy-Chpaman model fails to describe the EDL near the wall. Therefore, Stern proposed a new EDL model composed of two layers:

- The Gouy-Chpaman layer of diffuse ions (mobile)
- The Stern layer of absorbed ions (immobile)

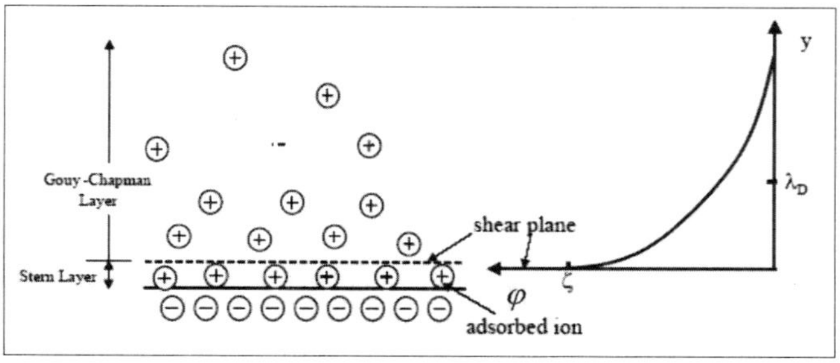


**Figure 4.1.4.** Gouy-Chapman Layer and Stern Layer.

Several characteristics of the Gouy-Chapman-Stern model are listed as the following:

- The immobile counter-ions ions are believed to be attracted to the charges of the wall by van der Waals forces
- The surface separating the 2 EDL layers is called the shear plane
- No-slip condition for the liquid motion is applied at the shear plane
- Zeta potential, $\zeta$, is defined as the potential value at the shear plane
- The Debye length, $\lambda_D$, is much larger than the Stern layer thickness

In conclusion, all the relations derived for the Gouy-Chapman model (diffuse layer) are valid in the (more physical) Stern model by substituting:

$$\varphi_w = \zeta \tag{4-12}$$

## 4.2 Electroosmotic Flow (EOF)

**Physical Model.** In a microchannel of which the walls are made of dielectric materials, the liquid inside the channel will move in the direction of the electric field applied between an upstream and downstream location. As shown in Figure 4.2.1, under constant electric field, the balance between the viscous and electric forces exerted on the migrating counter-ions in the mobile diffuse region of the EDL, at the solid-liquid interface, results in a steady liquid motion known as electroosmotic flow.

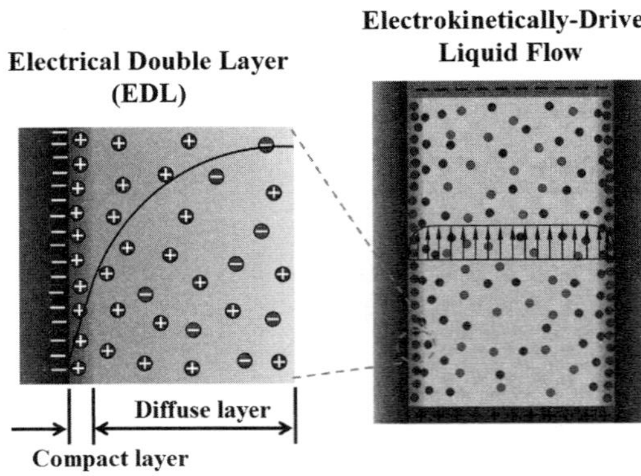


**Figure 4.2.1.** Electrical double layer and electrokinetically-driven flow.

**Mathematical Model.** The fluid motion in a microchannel, under an electric field, is governed by the incompressible continuity and Navier-Stokes equations:

$$\nabla \cdot \mathbf{v} = 0\,, \tag{4-13}$$

$$\rho\left(\frac{\partial \mathbf{v}}{\partial t} + (\mathbf{v} \cdot \nabla)\mathbf{v}\right) = -\nabla p + \mu \nabla^2 \mathbf{v} + \rho_e \mathbf{E}\,, \tag{4-14}$$

where $\rho$ and $\mu$ are the fluid mass density and dynamic viscosity, respectively, $\rho_e$ is the electric charge density, $\mathbf{E}$ is the electric field intensity vector, $p$ is the pressure and $\mathbf{v}$ is the velocity vector.

Under static or quasi-static electrostatic condition, the electric field vector is the negative gradient of the electric potential $\Phi$ which, in turn, is governed by the Poisson equation as follows:

$$\mathbf{E} = -\nabla \Phi \tag{4-15}$$

$$\nabla^2 \Phi = -\rho_e / \varepsilon \tag{4-16}$$

where $\varepsilon$ is the liquid permittivity. The distribution of the electric charge density in the liquid is the result of both the distribution of the surface charge at the channel walls and the externally applied electric field. The electric potential can then be decomposed into two components, the potential due to the applied electric field, $\phi$, and the potential due to the charge at the channel wall, $\varphi$, under the following conditions: (i) the Debye thickness is small compared with the channel diameter or height, (ii) the surface charge at the walls is not large, and (iii) the fluid velocity in the microchannel is too small to re-distribute the electric charge distribution (typical Reynolds number for electroosmotic flow is smaller than 0.01). Equation (4-16) can now be split into the following two equations:

$$\nabla^2 \phi = 0 \tag{4-17}$$

$$\nabla^2 \varphi = -\rho_e / \varepsilon \tag{4-18}$$

Utilizing the charge distribution proposed in Gouy-Chapman theory with linear approximation of the Debye-Huckel model (Bard & Faulkner, 1979), the EDL is modeled as a diffuse layer which consists of a series of laminae parallel to the micro channel wall. All the ions in these laminae are assumed to be in thermal equilibrium. Therefore, the ion distribution satisfies Boltzmann statistics. The electric potential at a fixed distance away from the channel wall is assumed to be a constant zeta potential $\zeta$. Hence, equation (4-18) can be simplified into:

$$\nabla^2 \varphi = \kappa^2 \varphi , \tag{4-19}$$

where $\kappa^{-1} = \lambda_D$ is the Debye length. It is important to note that there are many theories describing the electric double layer. Although the Gouy-Chapman theory and Debye-Huckel approximation include questionable assumptions and more complex theories have evolved, nonetheless it provides an often-used convenient model to describe the EDL.

**Electroosmotic flow in a microchannel.** To generate uniform electroomotic flow in a microchannel, a uniformly surface charged channel with a constant externally-applied electric field are required as illustrated in Figure 4.2.2.

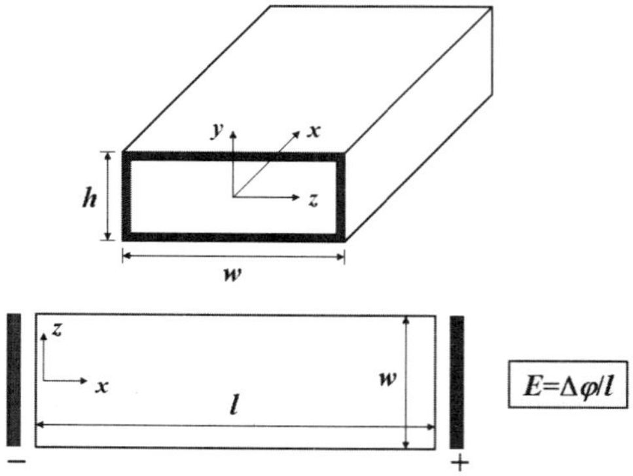


**Figure 4.2.2.** Electroosmotic flow in a microchannel.

For creeping flow in a microchannel ($Re << 1$), the momentum equation is simplified as:

$$0 = -\nabla p + \rho_e E + \mu \nabla^2 u \tag{4-20}$$

When combining with Poisson equation to electric-field can be further simplified to

$$-\nabla p - \varepsilon \varepsilon_0 E \nabla^2 \varphi + \mu \nabla^2 \mathrm{u} = 0 . \tag{4-21}$$

Assuming linear combination for the velocity, $u = u_E + u_P$, where $u_E$ is the velocity component from the electroosotic flow and $u_P$ is the Poiseuille flow due to the pressure gradient. The equation is expressed as:

$$-\nabla p - \varepsilon\varepsilon_0 E\nabla^2\varphi + \mu\nabla^2 u_E + \mu\nabla^2 u_P = 0 \tag{4-22}$$

By superposition, equation (3-22) is decomposed to

$$-\nabla p + \mu\nabla^2 u_P = 0\,, \tag{4-23}$$

in which the velocity field is due to the pressure force, and

$$-\varepsilon\varepsilon_0 E\nabla^2\varphi + \mu\nabla^2 u_E = 0\,, \tag{4-24}$$

in which electrokinetics is the driving force.

In the following result, only purely electrokinetically-driven flow (uniform pressure) will be considered because it is trivial to add the pressure gradient components to flow field by linear superposition, in which the governing equation is expressed as

$$\frac{d^2u_E}{dy^2} - \frac{\varepsilon\varepsilon_0 E}{\mu}\frac{d^2\varphi}{dy^2} = 0\,, \tag{4-25}$$

with to boundary conditions:

$$du_E / dy = d\varphi / dy = 0 \quad \text{at} \quad y = 0\ ; \tag{4-26a}$$

$$u_E = 0\,;\ \varphi = \zeta \quad \text{at} \quad y = h\ ; \tag{4-26b}$$

These conditions give solution of:

$$u_E(y) = \frac{\varepsilon\varepsilon_0 E\zeta}{\mu}\left[\frac{\cosh(\kappa y - \kappa h)}{\cosh \kappa h} - 1\right]. \tag{4-27}$$

The corresponding 1-D velocity distribution is shown in Figure 4.2.3.

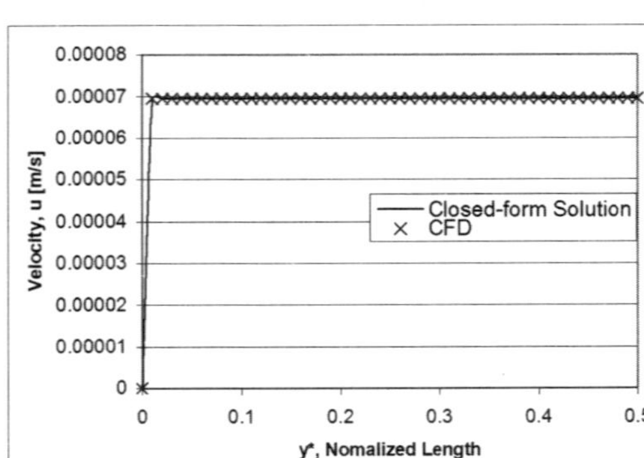


**Figure 4.2.3.** 2-D velocity profile of electroosmotic flow.

Under the thin EDL assumption (Probstein, 1994). $h/\lambda_D >> 100$, away from the wall, the solution (4-27) can further be simplified as

$$u_E(y) = -\frac{\varepsilon\varepsilon_0 E\zeta}{\mu}. \tag{4-28}$$

### 4.3 Complex Liquid Flows in Microchannels (EOF)

Microchannel flow is dominated by surface forces. Therefore, a very high pressure is required as the driving force. The flow is laminar and attached; it is very difficult to introduce turbulence and separation. However, selective patterning of the surface charges using standard photolithography techniques enables the generation of any arbitrary micro flow pattern using electrokinetics as the driving force. Using this technique, several vortex flow patterns are demonstrated.

**Out-of-plane vortex motion.** Under an externally applied streamwise electric field in the $x$-direction, the resulting electroosmotic flow is usually in the streamwise direction as well. In order to generate spanwise vorticity, all required is to induce non-uniform cross-stream distribution of the streamwise velocity component, i.e. $u(y)$. This can be achieved by patterning the top and bottom surfaces anti-symmetrically. Under a streamwise electric field, the liquid near the top and bottom surfaces flow in opposite directions giving rise to spanwise vorticity. The surface-charge pattern for generating a single out-of-plane spanwise vortex is shown in Figure 4.3.1(a). A positively or negatively charged zone is required on either the top or bottom surface, while the corresponding zone on the opposite surface carries the opposite charge. The active area is placed at the center, while the rest of the microchannel surface is kept neutral. The opposite electroosmotic slip, generated by a streamwise electric field, near the top and bottom surfaces leads to a re-circulating zone. The velocity field for this surface-charge configuration is plotted in Figure 4.3.1(b). A single out-of-plane vortex is clearly visible rotating clockwise around its axis in the spanwise $z$-direction.

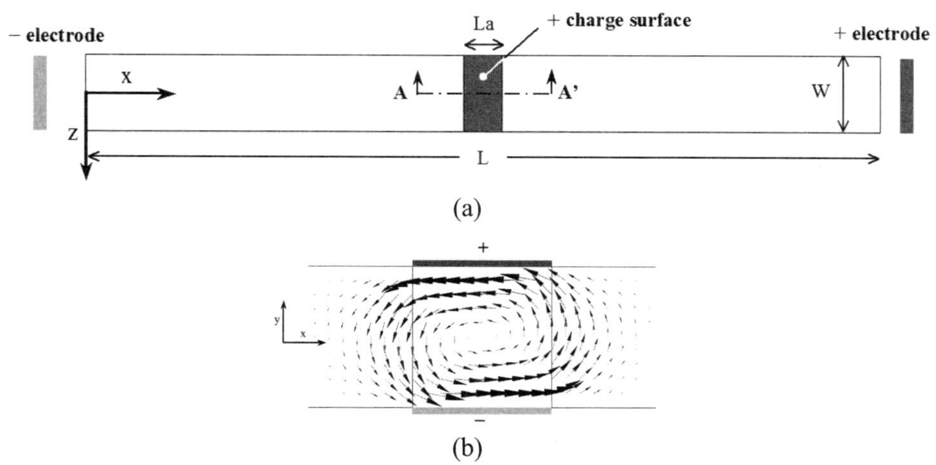


**Figure 4.3.1.** A single out-of-plane micro-vortex: (a) surface-charge pattern, and (b) the corresponding velocity field.

A train of co-rotating out-of-plane vortices can be formed by periodic duplication of the charge pattern required for a single vortex as schematically drawn in Figure 4.3.2(a) for 5 such cells. The corresponding velocity vector plot at the channel mid-plane is shown in Figure 4.3.2(b). Either the reversal of the electric field or flipping the charge pattern between the top and bottom surfaces will result in reversing the rotational direction of the vortex train.

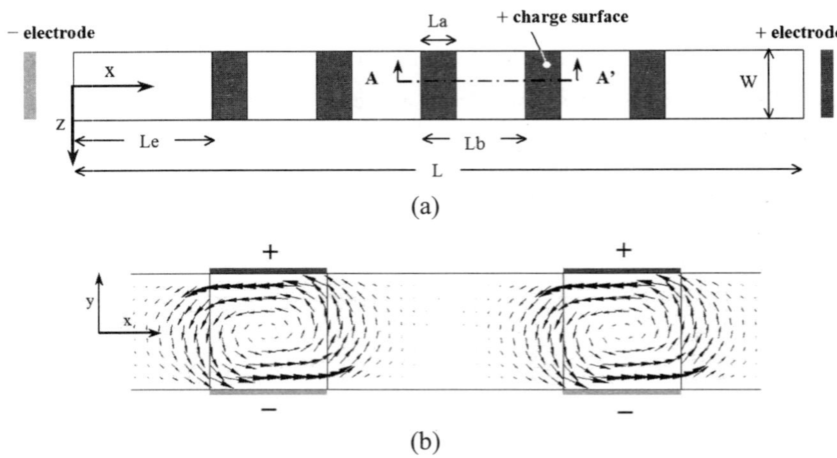


**Figure 4.3.2.** Pairs of out-of-plane co-rotating vortices: (a) surface-charge pattern, and (b) corresponding velocity field.

Counter-rotating pairs of out-of-plane vortices periodically distributed along the channel can be formed by flipping the charge pattern in every alternate cell in the configuration used for the co-rotating vortices as illustrated in Figure 4.3.3(a). The wavelength of the corresponding velocity vector field, depicted in Figure 4.3.3(b).

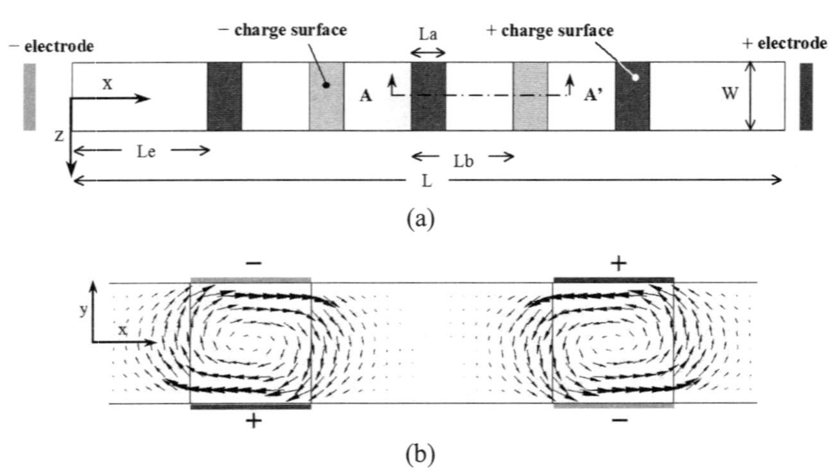


**Figure 4.3.3.** Pairs of out-of-plane counter-rotating vortices: (a) surface-charge pattern, and (b) corresponding velocity field.

Finally, these out-of-plane vortical flow fields are summarized using the spanwise vorticity component. The computed spanwise vorticity distribution along the microchannel at the channel centerline, $\Omega_z(x)$, is plotted in Figure 4.3.4 for all the liquid flows.

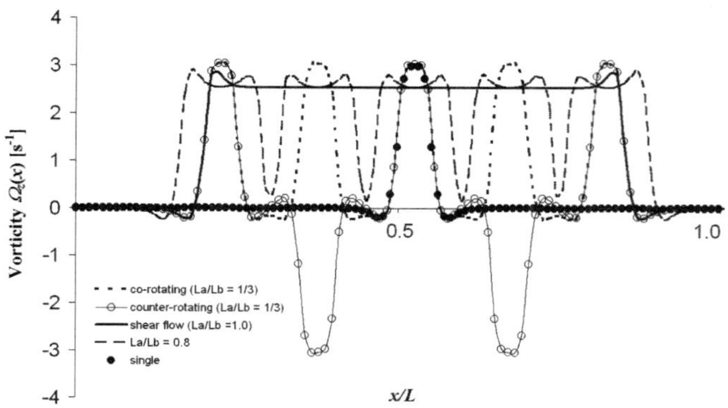


**Figure 4.3.4.** Spanwise vorticity distribution along the channel centerline, $\Omega_y(x)$, for the out-of-plane vortical flow fields.

**In-plane vortex motion.** A simple heterogeneous configuration to create a single in-plane vortex, as was first sketched by Ajdari (2002), is illustrated in Figure 4.3.5(a). A simple charge pattern of a positively charged zone next, in the spanwise direction, to a negatively charged zone is placed at equal distance from the channel inlet and outlet. The top and bottom surfaces are symmetrically charged, while the rest of the microchannel surface is neutral. The electro-osmotic slip generated by the streamwise electric field leads to a re-circulation zone. The resulting velocity vector at the channel mid-plane, parallel to the top and bottom surfaces, is plotted in Figure 4.3.5(b). A single in-plane vortex is obtained rotating clockwise.

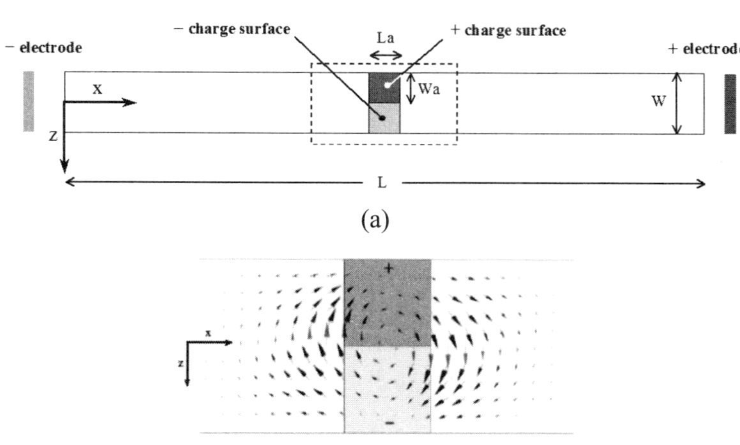


**Figure 4.3.5.** A single in-plane micro-vortex: (a) surface-charge pattern, and (b) the corresponding velocity field.

Streamwise periodic duplication of the charge pattern required for a single vortex is schematically drawn in Figure 4.3.6(a). A train of five in-plane co-rotating vortices is produced, of which two are plotted. All vortices rotate in the clockwise direction as shown in Figure 4.3.6(b). Again either reversal of the electric field or flipping the charge pattern will result in reversal of the rotational direction of the vortex train.

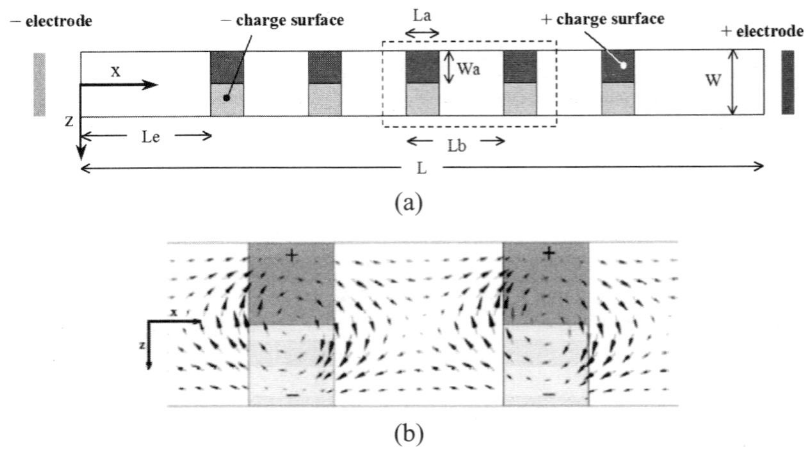


**Figure 4.3.6.** A train of in-plane co-rotating vortices: (a) surface-charge pattern, and (b) the corresponding velocity field.

It is also possible to generate a periodic flow of counter-rotating pairs of in-plane vortices. This can be done by flipping the charge pattern in every alternate cell in the configuration used for the co-rotating vortices as illustrated in Figure 4.3.7(a). The wavelength of the corresponding velocity vector field, shown in Figure 4.3.7(b).

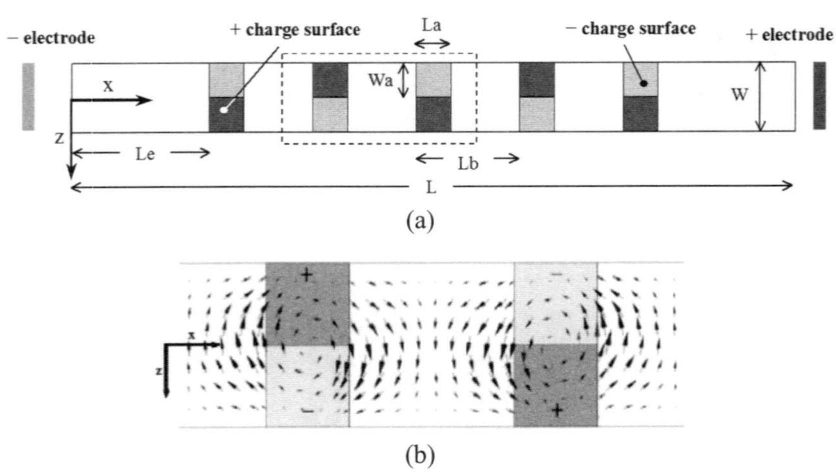


**Figure 4.3.7.** A train of in-plane counter-rotating vortices: (a) surface-charge pattern, and (b) the corresponding velocity field.

All these in-plane vortical flow fields can be summarized using the cross-stream vorticity component. The computed cross-stream vorticity distribution along the microchannel at the channel mid-height, $\Omega_y(x)$, is plotted in Figure 4.3.8.

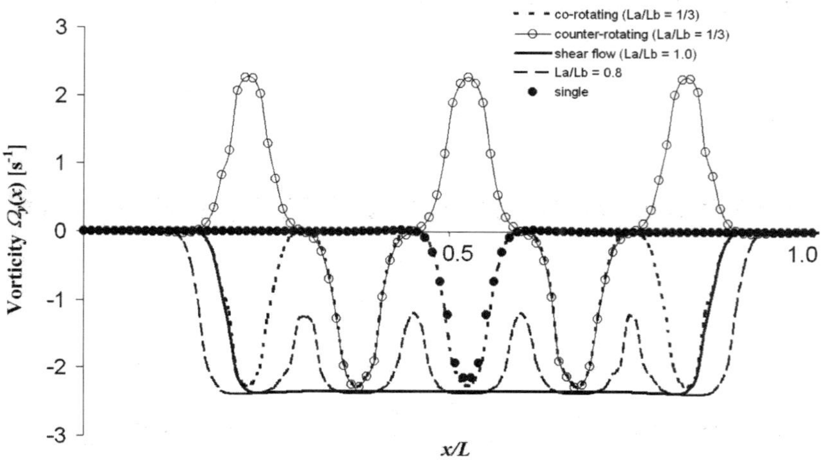


**Figure 4.3.8.** Cross-stream vorticity distribution along the channel centerline, $\Omega_z(x)$, for the in-plane vortical flow fields.

**Microscale liquid mixing.** Reduction in mixing time and length requires transverse velocity components of the main flow. A simple charge pattern, illustrated in Figure 4.3.9, is selected to generate in-plane vorticity. The pattern consists of three pairs of neighbouring positively and negatively charged regions. The negatively charged regions at the microchannel inlet and outlet are designed to provide the mean flow.

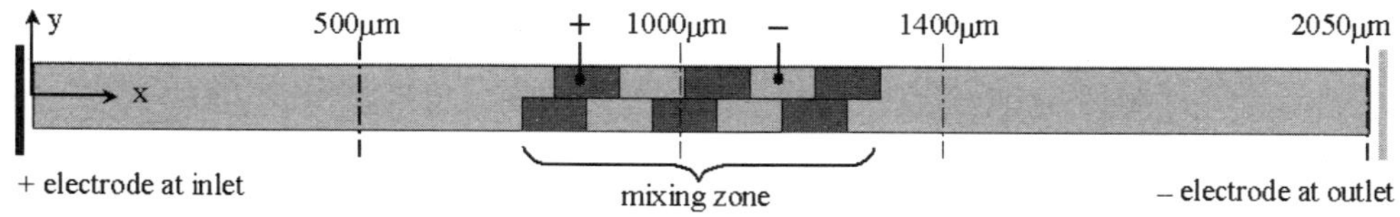


**Figure 4.3.9.** Plan view of the selected surface charge distribution along both the top and bottom surfaces.

Electric-field application across the channel results not only in electroosmotic flow driving the two liquids, but also in rapid mixing at the charge-pattern region as shown in Figure 4.3.10(a). In less than 1 min, the mixing process reaches a steady state with a well-mixed zone just downstream of the charge pattern shown in Figure 4.3.10(b). In comparison, the liquids in the electroosmotic flow developed in an identical channel with uniform negative charge, Figure 4.3.11, remain highly segregated.

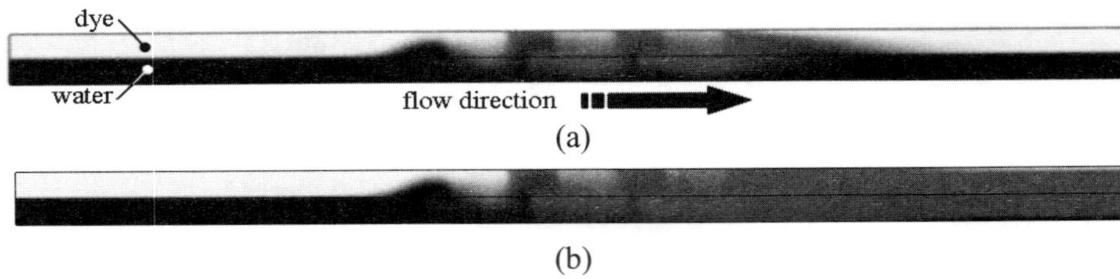


**Figure 4.3.10.** Planar concentration distribution of the two liquids in microchannel with the surface charge pattern in Fig. 4.3.9, (a) at transient state ($t$=5s) and (b) at steady state ($t$=50s).

**Figure 4.3.11.** Planar concentration distribution of the two liquids at steady state ($t$=50s) for a reference microchannel with uniform negative surface charge.

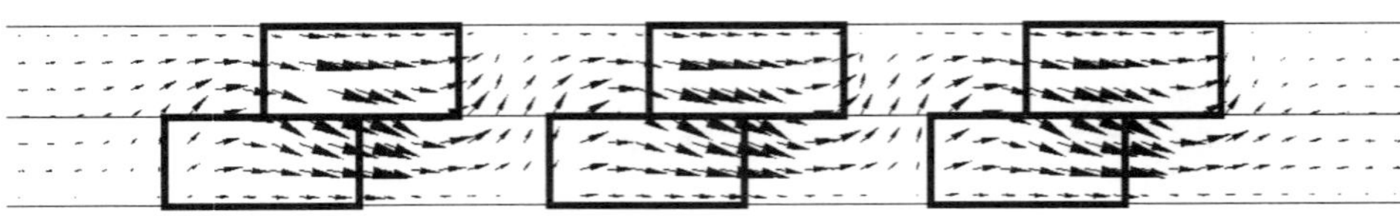

**Figure 4.3.12.** Mid-plane vector plot of the velocity distribution at the mixing zone.

The dramatically enhanced mixing in the microchannel with inhomegenous surface charge is due to the transverse velocity illustrated in Figure 4.3.12. Characterization of mixing process requires examination of the spanwise concentration profiles of the liquids as plotted in Figure 4.3.13 at the channel outlet for several time steps. Initially, the concentration increases from zero at half of the channel to unity at the other half. The concentration on both sides gradually approaches 0.5, indicating a well-mixed region, except near the sidewalls where the initial liquids are still present due to the no-slip boundary condition. Mixing can be measured by the relative liquid concentration; hence, a mixing index is defined as the ratio between the low and the high liquid concentration. Maximum mixing index of unity corresponds to 0.5 concentration of each liquid, while minimum mixing index of zero corresponds to complete absence of one of the liquids.

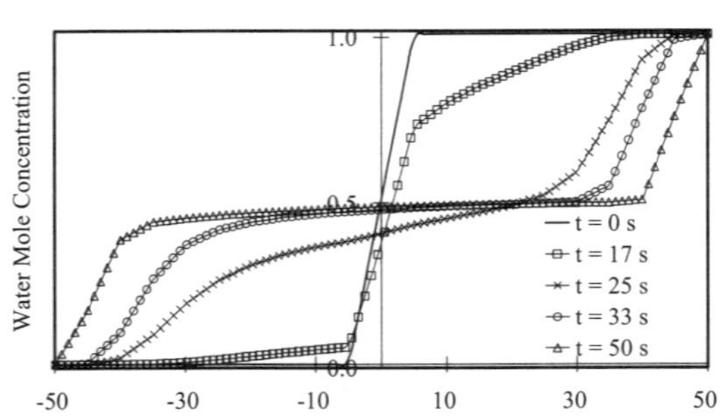

**Figure 4.3.13.** Spanwise water concentration profile at the channel outlet for several time steps.

The steady-state mixing profile across the channel mid-plane is shown in Figure 4.3.14 at several locations along the microchannel. Upstream the mixing zone, the mixing is poor except at the thin interface between the two liquids due to diffusion. The best mixing profile is obtained just downstream of the mixing zone where the index is close to one across the entire channel. The time evolution of the average mixing index at several cross-sections along the microchannel is plotted in Figure 4.3.15. The process is so rapid that within 30s, the index downstream of the mixing zone reaches its steady-state value of almost one.

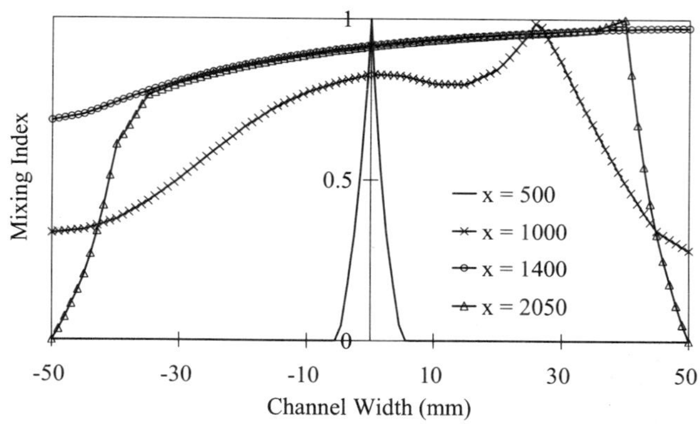


**Figure 4.3.14.** Mid-plane mixing-index profile at various locations along the channel at steady state ($t$=50s).

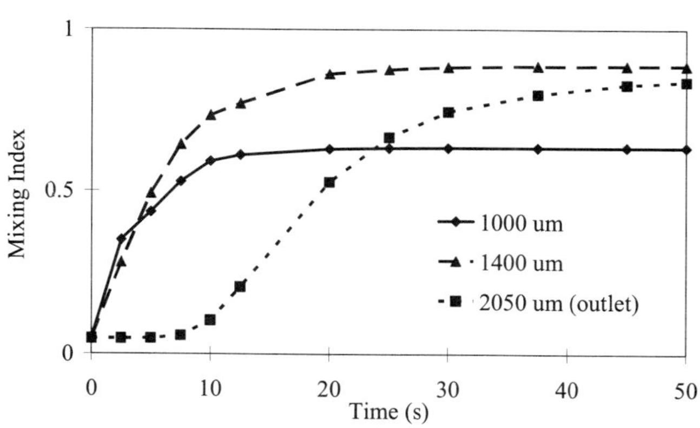


**Figure 4.3.15.** Time evolution of the averaged mixing index at several cross sections along the channel.

# 5 Fluidic Microsystems

## 5.1 Diagnostic techniques

**Mass flow rate.** In microscale gas flow, the mass flow rate inside a microchannel can be measured by traveling water meniscus in an outlet syringe (volume flow rate) as illustrated in Figure 5.1.1(a). One example of flow rate measurement with different working fluids is shown in Figure 5.1.1(b).

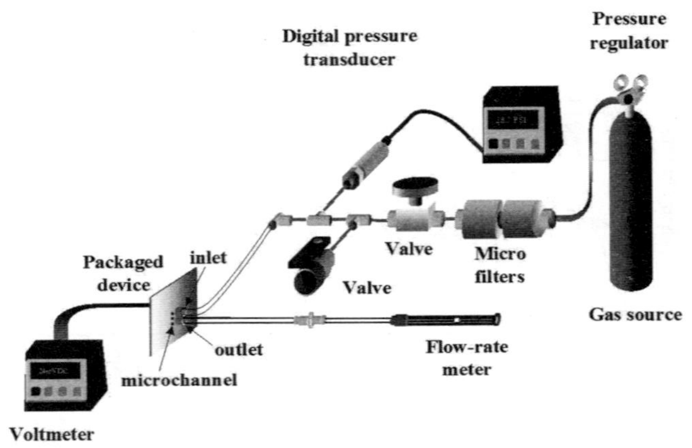


(a)

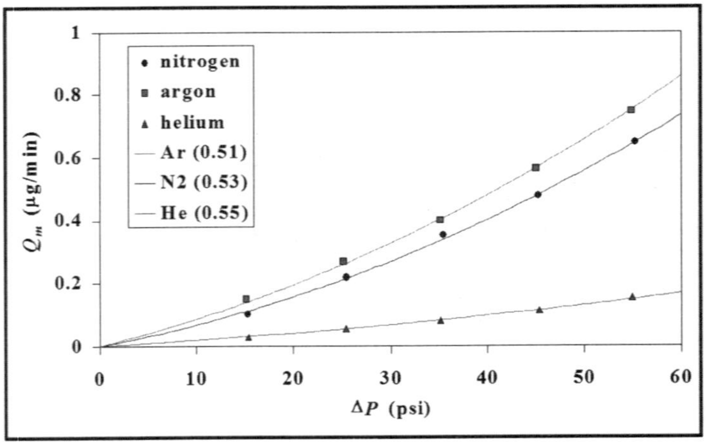


$L = 4000$ µm; $W = 40$ µm; $H = 0.53$ µm

(b)

**Figure 5.1.1.** Mass flow rate measurement (a) experimental setup and (b) of different working fluids (gases).

**Static pressure measurements.** Piezoresistive gage pressure sensors are integrated in a microchannel to measure the gas pressure distribution in Figure 5.1.2(a). The membrane constructed by thin film technology is deflected due to pressure difference across the microchannel. Resistance change is induced in the sensing element due to the resultant strain. The calibration curves of the pressure sensors and one case of measured streamwise pressure distribution are shown in Figure 5.1.2(b) and Figure 5.1.2(c) respectively.

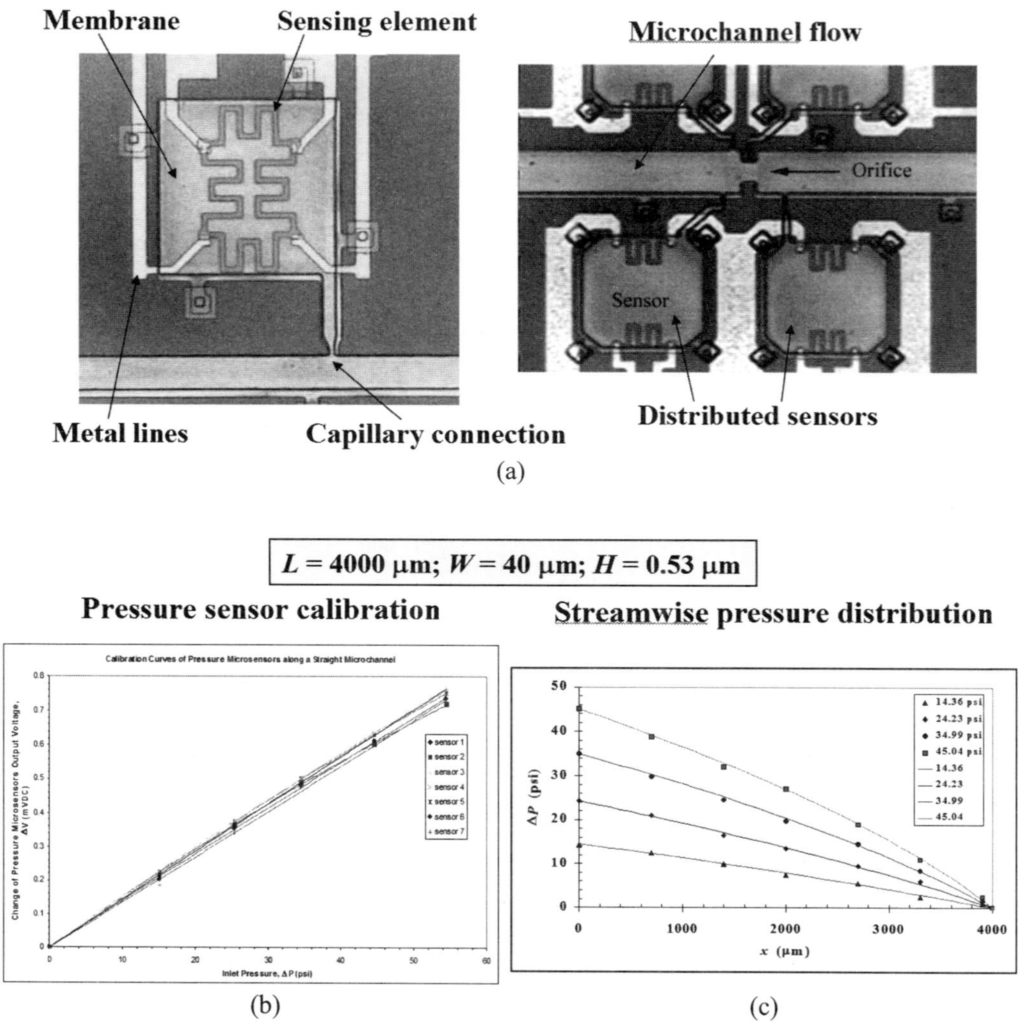


**Figure 5.1.2.** Static pressure measurements (a) by piezoresistive gage pressure sensors, (b) calibration curves and (c) streamwise pressure distribution.

**Surface shear stress.** Huang et al. (1999) built in a thermoresistive temperature sensor shown in Figure 5.1.3(a) to measure the surface stress of a moving fluid in microscale. The velocity gradient changes the temperature of sensing element; this temperature change in terms induces the change of the resistivity of the sensing element. The shear stress sensor calibration is also plotted in 5.1.3(b).

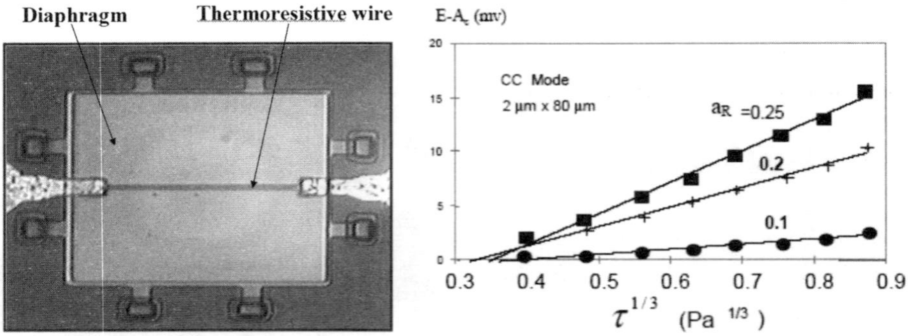


(a) (b)

**Figure 5.1.3.** Surface shear stress measurements (a) by thermoresistive temperature sensors and (b) sensor calibrations.

**Temperature distribution measurement.** Thermoresistive temperature sensor is fabricated in the microchannel Figure 5.1.4. Temperature change results in the change of resistance of the sensing element. The calibration curves of the thermoresistive sensors and one case of measured streamwise temperature distribution are plotted in Figure 5.1.5(a) and Figure 5.1.5(b) respectively.

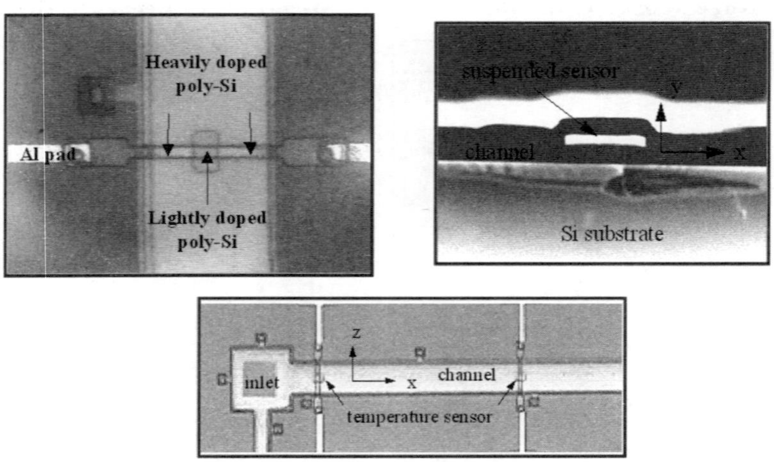


**Figure 5.1.4.** Integrated thermoresistive temperature sensors.

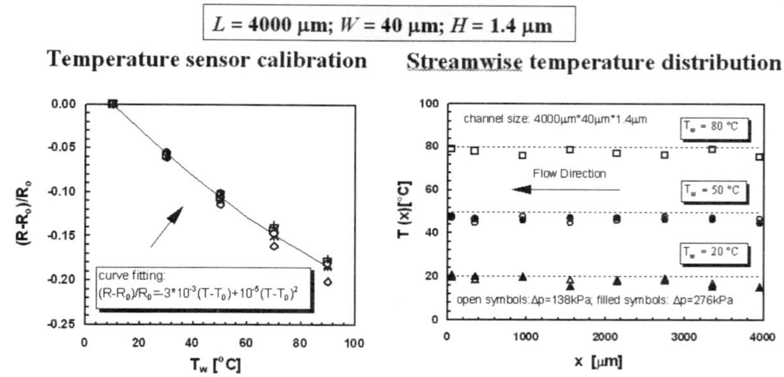


(a) (b)

**Figure 5.1.5.** Temperature distribution measurement: (a) calibration curves, and (b) streamwise temperature distribution.

**Flow visualization.** The flow field can be optically monitored under the microchannel using various techniques. Particles are laden in the working fluid in visualizing liquid flow in a micro-cavity in Figure 5.1.6(a). The phase change in convective boiling (vapor/liquid interface) and in a micromixer (liquid/gas interface) can be directly visualized under an optical microscope in Figure 5.1.6(b).

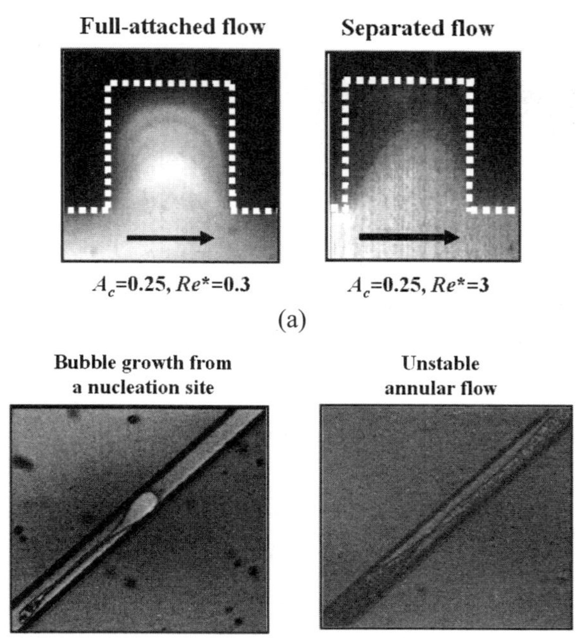


(b)

**Figure 5.1.6.** Flow visualizations: (a) liquid laden with solid, and (b) two-phase vapor/liquid flow in convective boiling.

**Velocity measurements (liquid).** Meinhart et al. (1999) developed micro particle image velocimetry (micro-PIV) technique to measure the flow field in micro scale. The seeded microspheres is tracked in space as a function of time. A typical experimental set-up is shown in Figure 5.1.7(a). The parabolic velocity profiles in microchannel liquid flow tracked by micro-PIV technique (Silber-Li et. al., 2003) are illustrated in Figure 5.1.7(b).

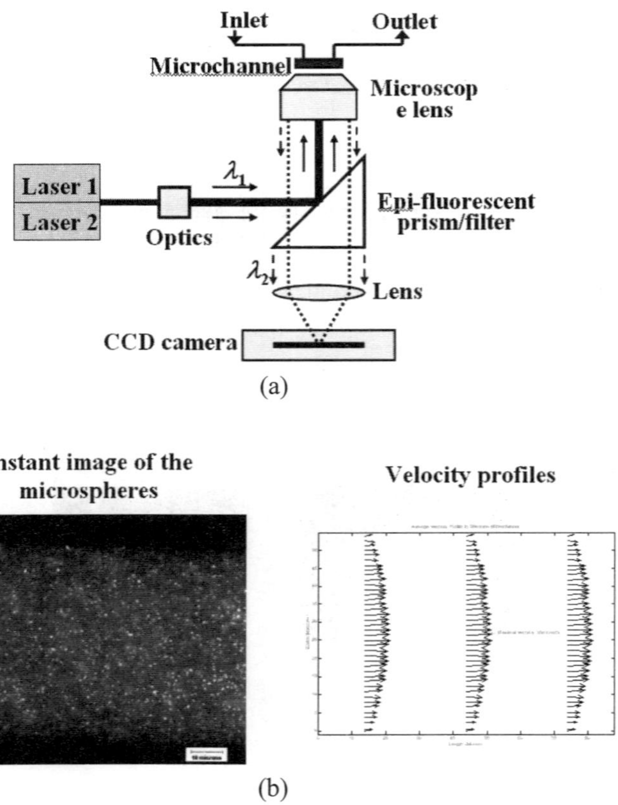


**Figure 5.1.7.** Velocity measurements (liquid) by micro particle image velocimetry (micro-PIV) technique (a) experimental setup and (b) velocity profiles by instant image of the microspheres.

## 5.2 Fabrication techniques

**Integrated micro systems for gas flow.** Using surface machining technique, a variety of integrated micro systems for gas flow are designed, fabricated and tested. The cross sections in the fabrication process are described in Figure 5.2.1(a). The different geometrical arrangements of microchannels for gas flow are summarized in Figure 5.2.1(b).

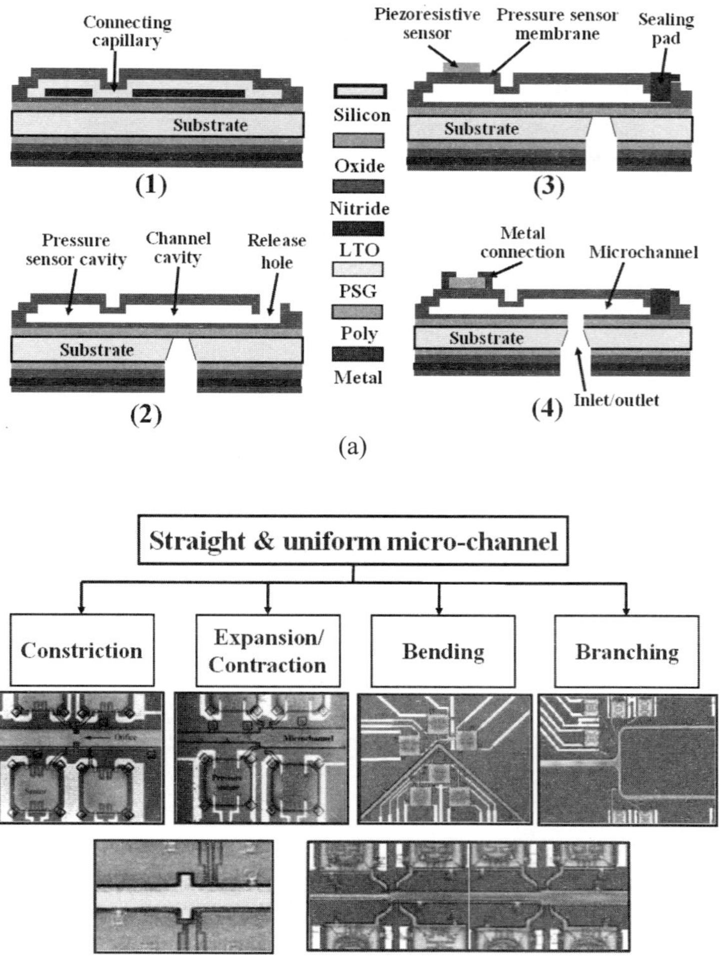


**Figure 5.2.1.** Integrated micro systems for gas flow (a) cross sections in the fabrication process and (b) summary of different geometric arrangements of fabricated devices.

**Surface modification.** Using surface chemistry combined with standard lift-off technique, micro-channel device which controls micro-scale liquid flow are fabricated and tested. The fabrication process begins with surface modification of the substrates to enhance the negative surface charge described in Figure 5.2.2(a). Electrostatic self-assembly method allows the polymer polyalylamine (PAH), in which the amine group ($-NH_3^+$) is positively charged, to be assembled to the substrate surface by electrostatic force in Figure 5.2.2(b). Arbitrary charged pattern arrangements can then be achieved by standard photolithography technique in lift-off process in Figure 5.2.2(c). The charge patterns can then be confirmed by excitation of PAH-fluorescein as illustrated in Figure 5.2.2(d). Experimental results showed that photoresist is a good masking material and PAH is stable in acetone and alcohol.

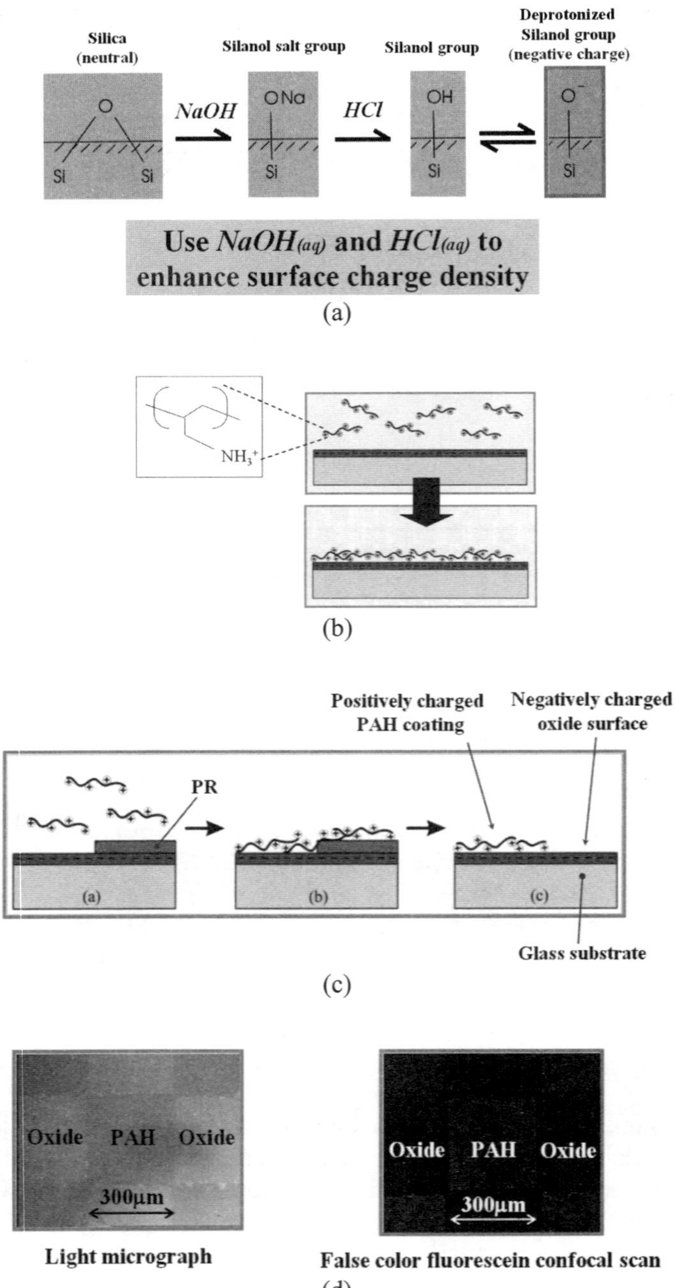


**Figure 5.2.2.** Surface modification by surface chemistry and lift-off technique (a) enhancement the negative surface charge, (b) electrostatic self-assembly of polymer polyalylamine (PAH), (c) standard photolithography technique in lift-off process and (d) PAH-fluorescein inspection.

**Integrated micro systems for liquid flow.** Using surface modification technique, various integrated micro systems for different microscale liquid flow patterns: bi-directional shear flow, out-of-plane and in-plane vortices and micro-mixing flow patterns are demonstrated. The fabrication process is described in Figure 5.2.3(a). Two microchannel devices are shown in Figure 5.2.3(b).

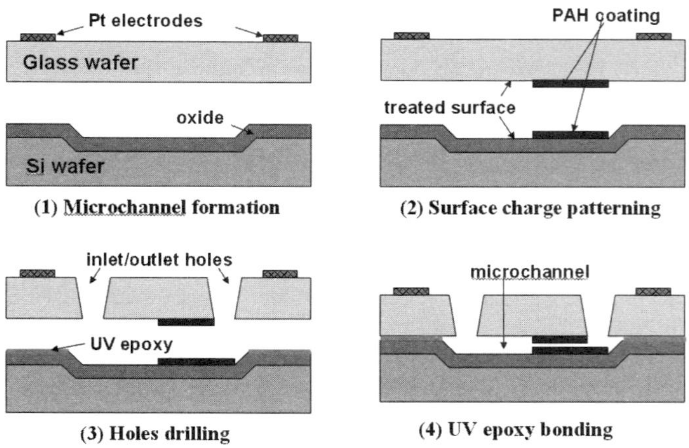


(a)

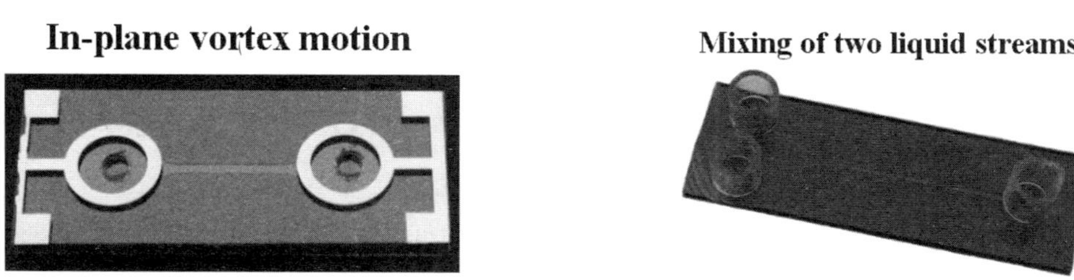


(b)

**Figure 5.2.3.** Integrated micro systems for liquid flow (a) cross sections in the fabrication process and (b) two examples of microchannel device.

## References

Acheson, D.J. (1995). *Elementary fluid dynamics*. Oxford University Press.

Ajdari, A. (2002). Transverse electrokinetic and microfluidic effects in micropatterned channels: lubrication analysis for slab geometries. *Physical Review E* 65 016301/1-9.

Arkilic, E.B., Schmidt, M.A. and Breuer, K.S. (1997). Gaseous slip flow in long microchannels, *J. Microelectromech. Syst.*, vol.6, pp.167-178.

Bard, A.J. and Faulkner, L.R. (1980). *Electrochemical Methods: Fundamentals and Applications*. New York, John Wiley & Sons.

Benedict R.P. (1980). *Fundamentals of Pipe Flow*. New-York, John Wiley & Sons.

Beskok, A. and Karniadakis, G.E. (1994). Simulation of heat and momentum transfer in complex micro-geometries, *J. Thermophys. Heat Transfer*, vol.8, pp.647-655.

Beskok, A., Karniadakis, G.E. and Trimmer, W. (1996). Rarefaction and compressibility effects in gas microflows. *J. Fluids Eng.* vol. 118, pp.448-456.

Bousse, L. et al. (2000). Electrokinetically controlled microfluidic analysis systems. *Annu. Rev. Biophys. Biomol. Struct.* vol. 29 pp.155-81.

Bradshaw, P. and Wong, F.Y.F. (1972). The reattachment and relaxation of a turbulent shear layer, *J. Fluid Mech.* vol. 52, pp. 113-135.

Burn, M. A., et al. (1998). An integrated nanoliter DNA analysis device. *Science* vol. 282 pp.484-487.

Cercignani, C., Illner, R. and Pulvirenti, M. (1994). *The Mathematical Theory of Dilute Gases*. Springer-Verlag.

Choi, S.B., Barron, R.F. and Warrington, R.O. (1991). Fluid flow and heat transfer in microtubes. *ASME DSC*, vol. 32, pp. 123-134.

Chow, V.T. (1959). *Open-channel hydraulics*. McGraw-Hill.

Chu, R.K.-H. and Zohar, Y. (2000). A class of discrete kinetic solutions for non-boundary-driven gas flow., *J. Non-Equilib. Thermodyn.* vol. 25, pp. 49-62.

Cochrane, T., Walters, K. and Webster, M.F. (1981). On Newtonian and non-Newtonian flow in complex geometries. *Phil. Trans. R. Soc. Lond.* A 301, pp. 163-181.

Davis, A.M.J. (1993). Periodic blocking in parallel shear or channel flow at low Reynolds number. *Physics of Fluids* A 5, pp. 800-809.

Dullien, F.A.L. and Azzam, M.I.S. (1973). Flow rate-pressure gradient measurements in periodically nonuniform capillary tubes. *AIChE, Journal* 19, pp. 222-9.

Devasenathipathy, S., Santiago, J. G. and Takehara, K. (2002). Particle tracking techniques for electrokinetic microchannel flows. *Anal. Chem.*, vol.74 pp. 3704-3713.

Dutta, P. and Beskok, A. (2001). Analytical solution of combined electroosmotic/pressure driven flows in two-dimensional straight channels: finite Debye layer effects. *Anal. Chem.*, vol.73 pp.1979-1986

Grossman, P.D. and Colburn J.C. (1992). *Capillary Electrophoresis: Theory and Practice*, Academic Press.

Harley J. C., Huang Y., Bau H. H. and Zemel J. N. (1995). Gas flow in micro-channels. *J. Fluid Mech.* vol. 284 pp. 257-274.

Hau, W.L.W., Trau, D.W., Sucher, N. J., Wong, M. and Zohar, Y. (2002). Micro flow patterns on demand using surface-chemistry technology. *Proc. 15th Int. Conf. Micro Electro Mechanical Systems* (MEMS'02) pp.475-78.

Hau, W.L.W., Lee, L.M., Lee, Y.K. and Zohar, Y. (2003). Electrically-driven vortical motion for mixing of liquids in a microchannel presented at the *7th Int. Conf. on Miniaturized Chemical and BioChemical Analysis Systems* (microTAS'03), *Squaw Valley*.

Hau, W.L.W., Lee, L.M., Lee, Y.K., Wong, M. and Zohar, Y. (2003). Experimental investigation of electrokinetically generated in-plane vorticity in a microchannel. *Proc. 12th Int. Conf. on Solid-State Sensors, Actuators and Microsystems* (Transducers'03) pp.651-54.

Hau, W.L.W., Trau, D.W., Sucher, N.J., Wong, M. and Zohar, Y. (2003). Surface-chemistry technology for microfluidics. *J. Micromech. Microeng.*, vol. 13, pp. 272-78

Ho, C.M. & Tai, Y.C. (1996). MEMS and its applications for flow control., *J. Fluids Eng*. vol. 118, pp. 437-447.

Ho, C.M., and Tai, Y.C. (1998). Micro-Electro-Mechanical Systems (MEMS) and Fluid Flows. *Ann. Rev. Fluid Mech*. vol. 30, pp. 579-612.

Huang, J.B., Jiang, F.K., Tai, Y.C. and Ho, C.M. (1999). A micro-electro-mechanical-system-based thermal shear-stress sensor with self-frequency compensation. *Exp. Fluids*, vol. 10, pp. 687-696.

Hunter, R.J. (1981). *Zeta Potential in Colliod Science: Principles and applications*. Academic Press.

Jiang, L., Wang, Y., Wong, M. and Zohar, Y. (1999). Fabrication and characterization of a microsystem for a microscale heat transfer study., *J. Micromech. Microeng*. vol. 9, pp. 422-428.

Karniadakis, G. and Beskok, A. (2002). *Micro flows: fundamentals and simulation*. Springer-Verlag

Kim, J.-H., Oztekin, A. and Neti, S. (2000). Instabilities in viscoelastic flow past a square cavity. *J. Non-Newtonian Fluid Mech*. vol. 90, pp. 261-81.

Lee, L.M., Hau, W.L.W., Lee, Y.-K. Wong, M. and Zohar, Y. (2005). Unsteady in-plane vortex motion in a microchannel liquid flow. *Proc. 18th Int. Micro Electro Mechanical Systems Conf.*, MEMS'05, pp. 734-737.

Lee, L.M., Hau, W.L.W., Lee, Y.-K. and Zohar, Y. (2005). Electrokinetically-driven in-plane vortical flows in a microchannel liquid flow using patterned surface charge. Submitted to *J. Fluid Mech.* for publication.

Lee, S.Y.K., Wong, M. and Zohar, Y. (2001). Gas flow in microchannels with bends. *J. Micromech. Microeng*. vol. 11, pp. 635-644.

Lee, S.Y.K., Wong, M. and Zohar, Y. (2001). Pressure losses in microchannels with bends. *Proc. 14th Int. Micro Electro Mechanical Systems Conf.*, MEMS'01, pp. 491-494.

Lee, S.Y.K., Wong, M. and Zohar, Y. (2001). Characterization of a mixing layer microdevice. T*he 11th Int. Conf. Solid-State Sensors and Actuators*, Transducers'01, Munich, Germany, June 2001.

Lee, S.Y.K., Yu, Z. T. F., Wong, M. and Zohar, Y. (2002). Gas flow in a microdevice with a mixing layer configuration. *J. Micromech. Microeng*. vol. 12, pp. 96-102.

Lee, W.Y., Lee, S.Y.K., Wong, M. and Zohar, Y. (2000). Microchannels in series with gradual contraction/expansion. *Proc. International Mechanical Engineering Congress & Exposition*, MEMS-Vol.2. pp. 467-472.

Lee, W.Y., Wong, M. and Zohar, Y. (2001). Flow separation in constriction microchannels, *Proc. 14th Int. Micro Electro Mechanical Systems Conf.*, MEMS'01, pp 495-498.

Lee, W. Y., Wong, M. and Zohar, Y. (2002). Microchannels in series connected via a contraction/expansion section. *J. Fluid Mech*. vol. 459, pp. 187-206.

Lee, W. Y., Wong, M. and Zohar, Y. (2002). Pressure loss in constriction microchannels. *J. Microelectromech. Syst*. Vol. 11, pp. 236-244.

Lee, Y.K., Lee, L.M., Hau, W.L.W. and Zohar, Y. (2004). Electrokinetically-driven out-of-plane vortices in a microchannel liquid flow using patterned surface charge. Submitted to *J. Fluid Mech.* for publication.

Li, X., Lee, W.Y., Wong, M. and Zohar, Y. (2000). Gas flow in constriction microdevices, *Sensors and Actuators A*, vol. 83, pp. 277-283.

Liggett, J. A. (1994). *Fluid mechanics*. McGraw-Hill.

Liu, J.Q., Tai, Y.C., Lee, J.J., Pong, K.C., Zohar, Y. and Ho, C.M. (1993). In situ monitoring and universal modeling of sacrificial PSG etching using hydrofluoric acid., *Proc. 6th IEEE Workshop on Micro Electro Mechanical Systems*, MEMS'93, pp. 71-76.

Liu, J.Q., Tai, Y.C., Pong, K.C. and Ho, C.M. (1993). Micro-machined channel/pressure sensor systems for micro flow studies. *Proc. of the 7th Int. Conf. on Solid-State Sensors and Actuators*, Transducers'93 (Japan) pp. 995-999.

Lord, R. (1974). Comparative measurements of tangential momentum and thermal accommodations on polished and on roughened steel spheres, In K. Karamcheti, eds *Rarefied Gas Dynamics*, Academic Press, New York.

Lukaszewicz, G. (1999). *Micropolar Fluids: Theory and Applications*, Birkhauser, Boston.

Mala, G.M., Li, D. and Dale, J.D. (1997). Heat transfer and fluid flow in microchannels, *Int. J. Heat Mass Transfer*, vol.40, pp.3079-3088.

Meinhart, C.D., Wereley, S.T. and Santiago, J.G. (1999). PIV measurement in a microchannel flow. *Exp. Fluids*, vol. 27, pp. 414-419.

Miller, D. S. (1990). *Internal flow systems.* BHR Group Limited.

Ng, A.S.W., Hau, W.L.H., Lee, Y.K. and Zohar, Y. (2004). Electrokinetic generation of microvortex patterns in a microchannel liquid flow. *J. Micromech. Microengng.*, vol. 14, pp. 247-255.

O'Brien, V. (1972). Closed streamlines associated with channel flow over a cavity. *The Physics of Fluids*, vol. 15, pp. 2089-97.

Panton, R. L. (1995). *Incompressible flow.* New York, John Wiley & Sons..

Pfahler, J., Harley, J.C., Bau, H. and Zemel, J.N. (1991). Gas and liquid flow in small channels. *ASME DSC*, vol. 32, pp. 49-60.

Piekos, E.S. and Breuer, K.S. (1996). Numerical modeling of micromechanical devices using the direct simulation Monte Carlo method. *J. Fluids Eng.* vol. 118, pp. 464-469.

Pong, K.C., Ho, C.M., Liu, J.Q. and Tai, Y.C. (1994). Non-linear pressure distribution in uniform microchannels. *ASME FED*, vol. 197, pp. 51-56.

Pozrikidis, C. (1999). *Little book of streamlines*. Academic Press.

Probstein, R.F. (1989). *Physiochemical Hydrodynamics.* Butterworths.

Prud'homme, R.K., Chapman, T.W. and Bowen, J.R. (1986). Laminar compressible flow in a tube. *Appl. Sci. Res.*, vol. 43, pp. 67-74.

Qian, S. and Bau, H. (2002). A Chaotic Electro-osmotic Stirrer. *Anal. Chem.,* vol. 74, pp. 3616-3625.

Reyes, D.R., Iossifidis, D., Auroux, P.A. and Manz, A. (2002). Micrototal Analysis Systems. 1. Introduction, Theory, and Technology. *Anal. Chem.* vol. 74(12), pp. 2623-2636.

Riegels, F. (1938). Z. Angew. *Math. Mech.* vol. 18, pp. 95-106.

Romero-Mendez, R., Sen, M., Yang, K. T. and McClain, R. (1999). Effect of fin spacing on convection in a plate fin and tube heat exchanger. *International Journal of Heat and Mass Transfer,* vol. 43, pp. 39-51.

Saad, M.A. (1993). *Compressible Fluid Flow*, 2nd ed. Prentice Hall.

Santiago, J.G., Wereley, S.T., Meinhart, C.D., Beebe, D.J. and Adrian, R.J. (1998). A particle image velocimetry system for microfluidics, *Exp. Fluids*, vol. 25, pp. 316-319.

Schaaf, S.A. and Chambre, P.L. (1961). *Flow of Rarefied Gases*, Princeton University Press, Princeton, New Jersey.

Schlichting, H. (1979). *Boundary Layer Theory*, 7th Ed., New York: McGraw-Hill.

Shen, C. and Floryan, J. M. (1985). Low Reynolds number flow over cavities. *Phys. Fluids*, vol. 28, pp. 3191.

Sherman, F. S. (1990). *Viscous flow*. McGraw-Hill.

Shih, J.C., Ho, C.M., Liu, J.Q. and Tai, Y.C. (1996). Monatomic and polyatomic gas flow through uniform microchannels. *ASME DSC.*, vol. 59, pp. 197-203.

Silber-Li, Z.H., Orlando, B. and Cui, H.H. (2003). Visualization of lower velocity flow fields in micro-channels. *7th. Asian Sym. Visualization*, Singapore, pp. 93-1.

Sreekanth, A.K. (1968). Slip flow through long circular tubes. in Rarefied Gas Dynamics (ed. L. Trilling & H.Y. Wchman). *Academic Press.*

Stroock, A.D., Weck, M., Chiu, D.T., Huck, W.T.S., Kenis, P.J.A., Ismagilov, R.F. and Whitesides, G. M. (2000). Patterning electro-osmotic flow with patterned surface charge. *Phys. Rev. Lett.* vol. 84, pp. 3314-17.

Stokes, V.K. (1984). Theories of Fluids with Microstructure, *Springer-Verlag, Berlin.*

Taneda, S. (1979). Visualization of separating Stokes flows. *J. of the Physical Society of Japan*, vol. 46, pp. 1935-42.

Taylor, G. I. (1971). A model for the boundary condition of a porous material. Part 1. *J. Fluid Mech.* vol. 49, pp. 319-26.

Van den Berg H.R., Seldam, C.A., and Gulik, P.S. (1993). Compressible laminar flow in a capillary. *J. Fluid Mech.*, vol. 246, pp. 1-20.

Van den Berg, H.R., ten Seldam, C.A. and Van der Gulik, P.S. (1993). Thermal effects in compressible viscous flow in a capillary. *Int. J. Thermophys.* vol. 14, pp. 865-892.

Wu, P. and Little, W.A. (1983). Measurements of friction factors for the flow of gases in very fine channels used for microminiature. *Joule-Thomson refrigerators. Cryogenics*, vol. 23, pp. 273-277.

Wu, S., Mai, J., Zohar, Y., Tai, Y.C. and Ho, C.M. (1998). A suspended microchannel with integrated temperature sensors for high-pressure flow studies, *Proc. 11th IEEE Workshop on Micro Electro Mechanical Systems*, MEMS'98. pp 87-92.

Yang, C., Li, D. and Masliyah, J.H. (1998). Modeling forced liquid convection in rectangular micro-channels with electrokinetic effects, *Int. J. Heat Mass Transfer*, vol.41, pp.4229–4249.

Zohar, Y., Lee, W. Y., Lee, S. Y. K., Jiang, L. and Tong, P. (2002). Subsonic gas flow in a straight and uniform microchannel. *J. Fluid Mech.* vol. 472, pp. 125-51.